Virtual Christianity

Virtual Christianity
Potential and Challenge for the Churches

Jean-Nicolas Bazin
and Jérôme Cottin

Risk
B O O K

WCC Publications, Geneva

Translated from the original French:
Vers un christianisme virtuel? Enjeux et défis d'Internet,
Geneva, Labor et Fides, 2003

Cover design: Marie Arnaud Snakkers
Cover photo: WCC/Peter Williams

ISBN 2-8254-1414-X

© 2004 WCC Publications
World Council of Churches
150 route de Ferney, P.O. Box 2100
1211 Geneva 2, Switzerland
Web site: http://www.wcc-coe.org

Published with the financial support of
World Association for Christian Communication
357 Kennington Lane
London SE11 5QY, United Kingdom
Web site: http://www.wacc.org.uk

No. 107 in the Risk Book Series

Printed in Switzerland

Table of Contents

Introduction

Rising to the challenge of virtual reality

Like all the institutions that make up our society, Christian churches are facing the challenge of responding to the emergence of a new reality – dubbed a virtual reality – yet Christianity is reluctant to embark on an uncharted journey into this new virtual world. This adventure into the unknown is offered – if not imposed – by "information and communication technologies", and more broadly by the online society that is emerging at the dawn of the 21st century. Should churches commit themselves to follow the path towards technological modernity?

It would be over-cautious for churches to refuse to venture into the virtual world at all: to do so would be a denial of the future, of progress and of the forward march of history, and this would not fit with the Christian vision of society which is characterized by faith in a future that is held in human hands, yet also in God's promise of a better and more just world. Refusing to face up to the technologies of the virtual world would contribute to the marginalization of the church and add to its own doubts about its role and vocation. We therefore have little choice but to move towards a meeting of these two realities, the spiritual and the virtual, and to take a gamble on that meeting being a harmonious one, as a result of shared objectives and a shared world.

We can, we think, move towards *e-Christianity* – that is, a Christianity that is in some way new, or renewed, by a thorough understanding of the new technologies: technologies that are no longer the technologies of the future, but are already the technologies of today. However, we must take care not to get drawn into an unrealistic longing for technological utopias, futuristic prophecies (fortune-telling?) or cheap fiction. Making a place for Christianity in the virtual world does not mean that Christianity should itself become virtual or disembodied, lacking a social or human grounding in reality. We must give some thought to the nature of virtual reality before we attempt to integrate it into a Christian vision of humanity and society.

Churches have taken a cautious, circumspect, even slightly fearful approach towards venturing into the virtual world. Are we really moving towards an e-Christianity and, if so, is that something to be welcomed?

A number of questions are posed by this. In this respect the churches have sometimes tended to echo the questions that others are asking, rather than suggesting answers that risk being invalidated by the very rapid evolution of new technologies in our society. Proximity to the emergence of these technologies and their effects on society makes it impossible to judge them objectively, and only tentative conclusions can be made. Moreover, churches have good reason to be cautious about this issue: because of their calling, they know that technology alone does not necessarily guarantee the authenticity of the message they seek to deliver. Technologies are invented by human beings, but the message that the churches must transmit – although it passes through human mediation – comes from God. Being too quick to employ new technologies may lead to the divine message being shaped or even substituted by a human medium. One might say that the more powerful and seductive the communication technologies, the more wary we should be of them, for it is when the image comes closest to resembling the reality it represents that it risks becoming an idol.

The experience of the global economy cannot but make us cautious. There was an unprecedented enthusiasm for technology start-ups and multimedia companies in the year 2000, then a collapse in the years that followed. Some people will see this as a confirmation of their fears, but others will see only the usual somersaults that accompany the advent of a new socio-economic phenomenon before regulations are put in place. Christianity cannot avoid virtual reality, but there is still a question concerning the existence of or desirability of an "e-Christianity".

The world of the Internet

This book will deal primarily with the Internet, since this is undoubtedly the most common and most popular manifestation of the virtual world. However, when talking about the Internet, it is not the technology behind it so much as the "world of the Internet" that will be discussed, because the Internet is more than just a technological medium.[1] Beyond its technological aspects, the Internet is indeed *a world*, a *new* world that is coming into being. In the space of just a few short years, it has already transformed our way of communicating, our language, our attitude to writing, our social relationships, our relationship with space and time, our way of learning and much more. The Internet is more than a new technology that we have to learn how to use – it is also the manifestation of that virtual world which both fascinates and frightens us.

However, there are two reasons why it is not enough to describe the "world" of the Internet simply as a "virtual" world. First, because a kind of "virtuality" has been a big part of the reality that we have lived and experienced for quite some time. The telephone, for instance, is a platform for virtual communication, yet nobody would dream of questioning its reality. Faith is another area in which we can talk about virtual realities, although of course in a different sense. God, the Holy Spirit, faith itself – aren't these "virtual" realities which are nonetheless perfectly real to believers?

Secondly, the virtual world of the Internet, when used properly and with circumspection, is not meant to be a substitute for reality: on the contrary, it gives greater depth to reality, intensifying, magnifying and amplifying it. Through the Internet we can escape into fantasy and explore imaginary worlds. We leave reality behind – and are as likely to find ourselves somewhere worse as somewhere better. But we can also return to familiar places, aspects of reality that make up the threads of our day-to-day world. The Internet adds a new dimension to our existing world rather than a new world in itself. This is why we have chosen to talk about "the world of the Internet", in the hope of avoiding giving the impression that this new phenomenon will create a completely new world, or new era. The "world of the Internet" seems an apt description of a complex phenomenon, of which the technological equipment involved is just one aspect.

Interested but cautious

The contemporary witness of the church and the calling of Christianity exist in juxtaposition with this world of the Internet. The church would no longer be the church if it did not witness to its faith in Jesus Christ. The church has a duty to turn outward and to be open to others; it must bear witness, and speak of the source of its life. This is its mission, its reason for being and its calling. Let us recall the words of Jesus to the first Christian community in Jerusalem, following his resurrection: "You will be my witnesses in Jerusalem, in all Judea and Samaria, and to the ends of the earth" (Acts 1:8b). Jesus precedes this commandment with the promise of a virtual reality to come: a power that is not physical or material, but is nonetheless real and leads to action in the world: the gift of the Holy Spirit. "But you will receive power when the Holy Spirit has come upon you" (Acts 1:8a).

However, we have to admit that, at present, our churches are not bearing witness very effectively. They are not entirely to blame: the church has little control over the external factors that are behind the declining interest in Christianity in the West. One of the many reasons for that decline is that we live in an information and communication society where technology has replaced people as bearers of news, and the image has replaced the word. Appearances have replaced reality and an infinite number of trivial, tautological messages have replaced the integrity of the word that once held meaning and a sense of transcendence. We believe that if the church wants to reverse, redirect or simply to resist this trend, a twofold approach is needed.

A Christian approach to the virtual world

First, the church should take a *practical* and *pragmatic* approach. It must make use of the new technologies in order to ensure that the Christian message is featured in society and the media. The church must make itself heard and seen if it wishes its message to reach as many people as possible. This is something that it will have to undertake beyond the circle of its followers. Just as the Reformation made use of the new technology of the printing press to spread its message, we need to do the same today, by using – without hesitation, but not without thinking – a technology that is as new and as universal in our time as printing was in the 16th century. This is not a question of giving in to fashion or to the ideology of efficiency, but simply of making the best possible use of the new possibilities afforded by technology in order to spread the news of the gospel. In addition to being a tool for information and communication, the Internet is also a potential tool for evangelization.

Secondly, the church needs to be willing to take an *idealistic* or even *humanistic* approach. It will have to make sure that its message does not find itself supplanted by the Internet's own message. Some apologists may say that the Internet has no message, but it would be more accurate to say that it has multiple messages – positive and negative ideologies based on how the Internet was invented, how it has grown, and how it is used. For example, three of the ideologies that pose difficulties for those who adhere to Christian values are:

- *A commercial ideology:* This has the goal of turning people into consumers – creating an Internet demand so that consuming takes precedence over using, and commerce take precedence over meaning.
- *A technological ideology:* This means that technology becomes an end in itself – i.e. functional communication prevails over normative communication, to use a distinction dear to media sociologists. The response to the omnipresence of the machine may bring a state of human solitude, as technology cannot entirely replace the human element which remains at the centre of life, of creation and of faith.
- *A religious ideology:* This is the most pernicious of the three, for it is the least perceptible and therefore the most difficult to spot; there is a religious ideology that is quietly at work in the world of the Internet which some people believe serves as a vehicle for anti-humanistic, syncretistic and pseudo-religious ideas. We will return to this issue later.

Thus, simply for the church to be able to continue to witness and fulfil its calling, and for human beings to remain human, we must find and then counter those ideologies that are contrary to the gospel. Making an emotive appeal to churches, Dominique Wolton, director of communication and information science and technology at the French National Centre for Scientific Research (CNRS), talks about fighting for the re-establishment of humanistic and Christian values in the field of communications in general, and in particular with regard to the new information and communication technologies. The new information and communication technologies tend to promote, openly or otherwise, values that are in conflict with the gospel: the cult of the superhuman; money as the only thing that matters; the loss of individual identity; confusion between creature and Creator; refusal to accept limitations of time and space; the absence of a sense of history; disembodiment; multiple personalities or multiple selves among users and much more.

Thinking about how we use the Internet
There will be no in-depth analysis of the origins of the Internet or of the technology behind it: others have already done this. Instead, this book will look at how the church can use the Internet, and we will do this in two ways, focusing on the relationship between the Internet and Christianity.

Chapter 1 begins by looking at the language and functions of the Internet and then sketches a process, more fully developed in appendix 1, for establishing an "Internet project" in a church. It concludes by showing how the Internet can be of use to the church globally and ecumenically, and how it has become a major tool for internal communication and for solidarity work within the World Council of Churches (WCC) in Geneva.

Chapter 2 examines how the new digital technologies are shaping *a new culture*, and that new culture is creating a new society. We will therefore engage in a dialogue with a number of Internet theorists, both those who are positive about new technologies, such as Patrice Flichy and Manuel

Castells, and those who, like Dominique Wolton and Philippe Breton, are more critical and who attack the ideological discourse and utopian views (including religious utopias) that have accompanied the arrival of these technologies.

Chapter 3 identifies five new issues raised by the Internet that will require vigilance on the part of the churches. There is a danger that the gospel message could be overtaken by the power of these new media: the church could, if it is not careful, find itself enslaved by these new technologies rather than the master of them.

But we must not stop there. All too often, the church has been content to criticize social phenomena without bothering to offer any specific, constructive proposals as to how to change things.[2]

The final chapter therefore examines ways of using the new possibilities that are opened up by the Internet for the benefit of the witness and the influence of the church, and addresses these as regards four areas in particular:

- *Reaching as many people as possible:* This new channel for information and communication can help Christianity fulfil its mission of witnessing to the power of the message of Jesus Christ everywhere, to as many people as possible, in attractive and accessible language.
- *Internal church communication and management:* The Internet allows church institutions to improve their internal communication by facilitating debate, formal statements and information-sharing.
- *Intensification and expression of faith:* The Internet can also assist in the expression of individual faith, by offering Christians powerful tools for living out and deepening their faith.
- *Ethical considerations:* The church has a responsibility to find a way of using the Internet differently from the rest of society. The church must make the Internet something that mitigates cultural and social inequalities, so that it will continue in the service of the poorest, of minority Christian communities and communities that are dis-

persed, and of a global Christianity which is now pre-
dominantly non-Western in nature.

NOTES

[1] We give a brief definition of the Internet in ch.1. In the context of this
discussion, we will make reference to the "Internet" both as a technological medium and as a social phenomenon. We consider the word
Internet as synonymous with new media or information and communication technologies, although we are aware that, from a technological point of view, these words describe different things.

[2] This is the critical principle that was applied to the theological dialec-
tique of Karl Barth in his "no" to natural theology: his "no" was not
followed by a "yes", i.e. by a specific proposal that would allow the
church to move forward in its cultural testimony and its social development. We should be prepared to welcome what is produced by individuals or by society and not just be suspicious of it.

1. The Internet and How to Use It

This chapter is intended as a practical introduction to the Internet, what it is and the vocabulary that goes with it. It is of course unnecessary for the experts, but experience has shown that accurate knowledge of the basic notions about the Internet is rather limited. What follows is therefore intended to help the reader better understand the subsequent chapters.

What *is* the Internet?

The Internet is a network of local networks that makes it possible to convey information from one computer to another, and from one network to another.

A computer – which is necessary in order to use the Internet – is connected to the worldwide network, to which millions of other computers are also connected. The system boasts exceptional technical reliability. It was initially conceived in the 1960s, by the American Advanced Research Project Agency (ARPA) of the US Department of Defence. The network, called ARPANET, was designed to be capable of withstanding a nuclear attack, its structure in the form of a vast "net" or "spider's web" ensuring that, if one part of the net was destroyed, the rest of the mesh would allow the network to continue functioning.

This network came to be widely used in universities, at first mainly to exchange e-mails. Then, in the 1980s, the name "Internet" emerged, with a variety of functions such as discussion groups, and then eventually "Internet" became synonymous with the "Web". It was during this period that an "Internet culture" based on the free sharing of resources took shape, and that some of the ethical rules that Internet users impose on themselves ("netiquette") took hold.

The 1990s saw the use of the Internet explode in all spheres of society, by commercial companies, political institutions, community organizations – and churches.

The Internet fits within the broader technological context of what are known collectively as "new media" or information and communication technologies – communications media that have emerged as a result of the convergence and interconnection of computer, telecommunications and audiovisual technologies which are now all interconnected. That convergence is made possible by the digitalization of information. The platforms for these new media (computer, telephone, television) are "networked", but information and communication technologies also include stand-alone media, such as CD-ROMs.

The new technologies have applications in four very different areas: work, leisure, provision of services and education. These four areas directly concern Christianity, because the church is simultaneously a place of work, a service provider, an organization that interacts with people during their free time and one that has very specific educational goals.

There are three words that sum up why we must get to know the Internet: speed, popularity, efficiency.

Speed refers to the rapidity of communication, because we can now communicate "in real time" over great distances. In electronic exchanges, distance is no longer a factor. It also refers to the rapid popularization of the Internet, which in the space of a few short years has gone from being something that was initially used only by scientists and the military to something that we no longer hesitate to describe as a mass-market product, comparable to the telephone, car or television. Business has played an important role in this, to the point that technological innovation has been given priority over educational considerations and the development of the infrastructure to make electronic communication possible.

Popularity is the second characteristic. The Internet is now an established medium, accessible to a large number of families and individuals. The number of users is increasing so fast that any statistics given here will be immediately out of date. The World Wide Web, one of the first mass applications of the Net, dates from just 1990. Fourteen years later, the Internet was used by 55 percent of the population of the USA and 23 percent of Europeans. There are still inequalities, of course, and the risk of a "digital divide" is something that we will come back to later, but those inequalities are much less pronounced than in other areas (education, housing, social status, and so on), because of the relatively low cost of computer equipment and of connecting networks and technologies.

Efficiency: The Internet responds to all kinds of demands. Its efficiency is due not only to the number of services it offers, but is also a result of its speed – in an instant we can be connected with a place or a service on the other side of the world.

Dominique Wolton identifies three coexisting functions of the Internet that combine technological performance with the human and social capacity for communication:

1. The Internet delivers *information:* information which is necessarily both varied and abundant for today's complex society.
2. The Internet promotes *expression:* we can say and tell one another all sorts of things. The Internet reflects and encourages our need to talk to one another in a society which is free but in which people are more and more alone.
3. The Internet is a forum for *communication:* this recognizes the difficulty we have in understanding one another.

WAYS OF USING THE INTERNET

The Internet gives us the opportunity to communicate with people who are a long way away. We can think of it as a more evolved form of the telephone, and it uses telephone technology as one of its basic support systems.

The following list is intended to explain certain concepts that can sometimes still seem vague.

E-mail (electronic mail)

This is one of the most popular applications of the Internet: e-mail allows you to write a message and send it to the "electronic mail box" of the addressee, who can then see a list of the messages that he or she has received on his or her computer screen. The recipient can read them, print them or send them on to another Internet user.

One of the advantages of e-mail is that it is easy to learn and easy to use.

Another advantage of e-mail (compared with a letter) is that it is quick to write (you no longer need to hunt for paper and an envelope, look up the address and post the letter) and it is also quickly delivered (between a few seconds and half a day to travel around the globe).

A third advantage of e-mail (compared with the telephone) is that the person with whom you are communicating does not need to be present at the time. We send a message when it is convenient for us and recipients read it when it is convenient for them; this avoids the problem of disturbing people with a phone call at an inconvenient time, or having to try several times to get hold of someone. E-mail is also a very effective way of overcoming the problem of time differences, for example when people being addressed are in the United States, Europe and the Philippines.

Mailing lists

A mailing list allows you to compile lists of addresses in order to be able to send a message to multiple recipients simultaneously; for example, the news agency Ecumenical News International uses a mailing list to send its releases out to many journalists at the same time.

Web pages

Web pages are another popular application of the Internet. They make it possible for users to view, on their own computer screens, a page made up of text and images that is located on another computer. For example, a parish can publish a Web page showing the times of services which may be consulted by anybody with access to the Internet.

A function called "hypertext links" or "hyperlinks" makes it possible to move easily from one page to another by using the mouse to click on a link word. This makes it possible to compile several pages on a subject; a set of pages of this kind is called a "website".

Using hyperlinks to move from page to page and from website to website is called "surfing" the Web.

Portals

A portal is a website that specializes in searching for Web pages. It is essentially a directory of the Web pages that can be found in a predetermined category (economics, religion, technology, etc.). Of course, given the massive number of pages that exist on the Internet, portals will only reference the most important ones, but even this can be very useful. There are both "generalist" portals (of which the best known is undoubtedly *www.yahoo.com*), and specialist portals, such as in the US or in Europe, *www.churchnet.org.uk, www.protestants.org* (the portal for French Protestantism) or *www.protestant.ch* (the Protestant portal for French-speaking Switzerland).

Search engines

A search engine is a tool that allows users to search web pages by typing in "key words" – *Calvin*, for example. The search engine then scans all the pages that it knows about, looking for the word *Calvin*. Although search engines are extremely powerful, it takes very little experience to be able to use them to find very quickly just what you are looking for. A search for the word *Calvin* using the search engine google.com finds around 5,330,000 web pages where the word is mentioned. In fact, it is advisable to narrow down the search by being more specific about what it is you are looking for: for example, a search for the words *John Calvin* produces 1,710,000 pages, while *"John Calvin"* with quotation marks produces 143,000 pages[1] – and so on, depending on what you are looking for. The best known search engine is www.google.com

Discussion forums

Discussion forums make it possible for several people to engage in dialogue or conversation on an agreed topic. One person asks a question, for instance, "What should our attitude be towards the use of new technologies by the church?" Anybody on-line may suggest an answer and also may comment on other people's answers. It is a public debate, because

14

everybody who has access to the Internet can read it and take part in it. This type of dialogue does not take place instantaneously – various people might contribute to it after an interval of several hours or even several days (an example of asynchronous communication).

Chat rooms and instant messaging

This is the equivalent of a face-to-face conversation via the Internet that allows people to communicate "live" with other Internet users by sending written messages that result in an interactive conversation.

Teleconferencing

Teleconferencing is the audiovisual version of "instant messaging"; it makes it possible for two people to share files remotely (for example, to work together on a Microsoft Word document that is visible on each of their screens), and even to hear and see each other. The technology to do this is available very cheaply, but one should not expect high quality sound or pictures. It is possible to obtain higher quality resolution using specialized equipment but with a price tag to match, or using sound only (as the visual image uses a much greater quantity of the communication capacity).

It is sometimes advisable to combine a low-budget solution to allow several people to work on the same text while talking on the telephone (examples: Yahoo Messenger, MSN Messenger, Skype).

Work groups and shared documents

Shared documents are stored in a directory in such a way that the information is accessible to more than one user. Work groups are also an opportunity for "information sharing". A document produced by one person is made accessible electronically to other people.

Imagine that a journalist on a church newspaper writes an article. Once finished, the article is automatically made available to the chief editor who then reads the article and gives approval. This then makes the article available to the editorial team in charge of the paper's composition.

This application is very useful and setting it up requires a degree of experience and a methodical approach.

WHAT DOES THE INTERNET HAVE TO OFFER THAT IS NEW?

The arrival of the Internet has provoked a paradigm shift comparable to the shift that accompanied the invention of the printing press in the 16th century, away from a hand-written literary culture and towards duplication and the wide dissemination of knowledge.

The Internet has taken that phenomenon to new levels, as a means for sharing information at a previously unknown rate in terms of the universality, coverage, ease of use and wealth of information.

Universality

The Internet is a real symbol of *universality*; it reduces distances and makes our planet seem like a small yet global village. The Internet is also a very powerful vehicle for the globalization of the exchange of information. Churches, which have a universal message to convey, have much to gain from this aspect of the Internet. However, the universalization of culture brings up the issue of globalization; we are facing both opportunities and challenges that are unprecedented.

Decentralized power

The Internet is a world that eludes any central control; neither governments nor big commercial companies can pin it down. It is a symbol of the total decentralization of power. Complete freedom of expression prevails – which is good in some respects, but also means that the most unpleasant ideas can circulate unrestricted. We will look at this aspect in more detail later. Decentralization also gives Internet users themselves the opportunity to think about the code of ethics applied to the Internet – "netiquette" – aiming to and encouraging Internet users to behave ethically.

The Internet is an extremely democratic environment: access to it is affordable, and everyone – from ordinary citizens to heads of state – is on an equal footing.

In businesses or modern organizations that are structured along the lines of a network of people rather than as a hierarchical structure, the Internet offers very effective horizontal communication. The centre is no longer there to control, but to moderate in order to ensure that the system is working properly. This characteristic clearly makes the Internet a powerful tool for decentralized organizations (which some may see as being more in line with Christian or at least Protestant culture). This can make the Internet a threat for those who are determined to control everything.

A new world

The Internet represents a new world, a new culture. The conquest of the Internet is a little reminiscent of the conquest of a new world (as the Americas once were), both in the dash for quick fortunes, the "start-ups" followed by the crushing reversal of fortunes and the presence of highway robbers waiting to send a virus or to steal a credit card number.

Of course, it is important to remain rational and to understand that we must look at this new dimension of our world from the point of view of our Christian values. Some religious bodies have already taken up positions in this new media landscape by producing very attractive websites. The problem is that it is often difficult to tell who is hiding behind the attractive website that is open "to all who believe in the Lord Jesus Christ". There are even "cyber churches", which are communities made up of scattered people who know each other only via the Net!

A new way of giving and receiving

The notion of "something for free" is one value that has emerged as part of the Internet culture, and it has posed a serious challenge to the business world. The Internet may be the home of business, but it is also the home of freely sharing. A phenomenal amount of knowledge, in the form of directories, guidebooks, encyclopedias, reviews, versions of the Bible and software is available on the Internet free of charge. It is becoming increasingly difficult, if not futile, to try to charge for information.

People are waking up to the fact that by opening up and sharing what we have for free, we can all be better off. Think about the amount of knowledge that we have, individually or within our organizations, that could benefit a neighbour, another parishioner, another church, a far-off country. The opportunities afforded by the Internet challenge us to serve God as best we can and to put our knowledge at the service of our neighbours.

The flipside of this is that the Internet becomes an arena for the unscrupulous ripping off of copyright (software, images, music, video).

THE CHALLENGES

The advent of the Internet raises a number of challenges to which we need to respond by defining a framework for the appropriate and equitable use of these new means of communication.

Virtual reality

The notion of virtual reality has its origins in the invention of computer games that use a "virtual" reality. Players wear a screen that masks their vision and projects the image of a landscape created by a computer and a special mechanical glove that allows them to see and to touch an object produced by the computer and thus to create a virtual world where artificial objects may be touched as if they were real.

All these tools allow us to be together or, more accurately, to be together *virtually*. The word "virtual" is used because people are not physically next to each other, but the technology that connects them creates a sense of proximity, as the telephone once did.

Despite its more trivial side, "virtual reality" can become very complicated. For example, sending an e-mail message is more of a virtual experience than communicating on the telephone, and if a computer then replies automatically to the message the exchange becomes entirely virtual. Indeed, some organizations that are overwhelmed by floods of e-mails have developed software that can automatically decipher the key points of an e-mail message and compose a "personalized" response.

The trick is being able to tell the difference between what is virtual and what is actual when you are using the Internet. We will consider this dialectic between the real and the virtual later on.

Escapism

Some people find that, having spent many hours navigating their way around "the world of the Internet", they feel at home there. There are a number of reasons for this.

First, using a computer can become a real pleasure in itself. Quite apart from the services that it provides, there is pleasure in discovering new skills, in creating, in mastering the challenges set by the machine. For some people who have difficulty communicating or who are experiencing problems in their everyday lives, it can be a comfort to immerse themselves in a private tête-à-tête with the computer, because the computer offers a world so logical that there are no relationship problems, and the challenges that are made are not threats but just equations that can quickly be solved.

Who hasn't intended to spend 15 minutes writing a memo and ended up spending an hour, having become enthralled by the engagement of creative imagination and the pleasure of seeing work take shape from initial disorder? Who hasn't come to the end of that hour with a guilty feeling of having become carried away, in addition to the sensation of pleasure?

Secondly, the world of the Internet is a convivial, anonymous place with relatively simple relationship rules; it is a fascinating world, both futuristic and human, and a place that is so user-friendly and so rational that seeking refuge there can be tempting.

The Internet (like reading or television) can therefore, if you are not careful, be addictive; it may become an environment that can serve as an escape from the real world.

A culture of intuition

For years, the media have been making us aware of our intuitive side, as opposed to our rational side. Although we may like a poem or a picture, we may not necessarily know why we like it; nevertheless we come to accept its message, the values that it conveys, but without having analyzed them. This is the mechanism used by advertisers (using a seductive message to create an instinctive, non-reasoned, emotive affiliation). Television also invites us to live in the moment.

The Internet leads us right into this intuitive world. Websites are first of all places that we visit briefly; visitors spend only a short time on a page, and the information must be available in "a few mouse clicks" or most users will become lost or bored.

E-mail does allow for some detachment simply because it is a written document that takes a certain amount of time to type. But the quantity of e-mails we receive encourages us to write quickly, and to "cut to the chase". The speed of the technology nudges us towards exchanges that are less considered and more spontaneous. Once printed, a message changes; suddenly, it looks like an official document. A great many misunderstandings arise out of the discrepancy between the spontaneous way in which e-mails are drafted and recipients' perception of the result as a written document.

The culture of the intuitive can seem destabilizing to those traditions of Christianity that value intellectual rigour. Other more charismatic currents find that they can express themselves more naturally through the spontaneity of the Internet. However, these two ways of approaching the world are complementary. We do need to accept and strengthen our intuitive side, because it is our intuition that is the vehicle for our capacity for spontaneous judgment. It is also important, though, to make sure that we do not allow ourselves to get carried away by that tendency; rather, we should come to understand clearly how to use these new tools properly.

The challenge of an Internet project for the church

It is very difficult for a church to embark on a project involving new technologies for the simple reason that there is in a given church only limited knowledge, if there is any technical knowledge at all.

The first thing to do is to find professionals who understand both worlds and who can bridge the world of technicians and the world of the church.

Appendix 1 explains the basic notions regarding an Internet project, proposes an approach in stages and some key role descriptions to ensure the success of the project.

It is also important to notice that for cultural reasons, any project will need to last some time and occur in small phases, in order to allow a progressive appropriation of this new world by the whole church.

Internet access and the digital divide

One of the problems created by the Internet is the "digital divide", a term used to describe the fact that the Internet is far from universal; some people may have access to it, but others do not. The digital divide is what separates these two categories of people.

The first main reason for the digital divide could be described as cultural: because these technologies are relatively new, many people do not realize their potential, do not know how to use them or are perhaps just afraid to try.

The second reason is technological; poor telephone coverage in developing countries quite simply puts the Internet completely out of reach of the majority of the population.

Here are the statistics for the year 2003, provided by the International Telecommunications Union (ITU) in Geneva:

Population   Phones

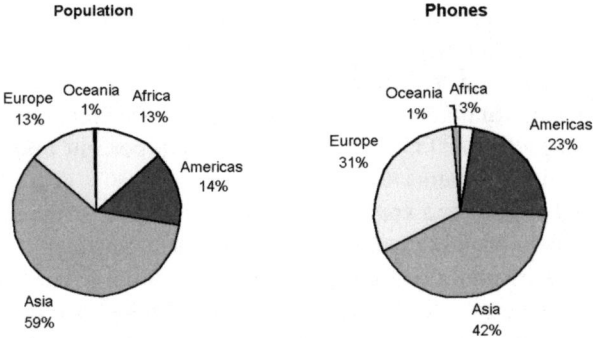

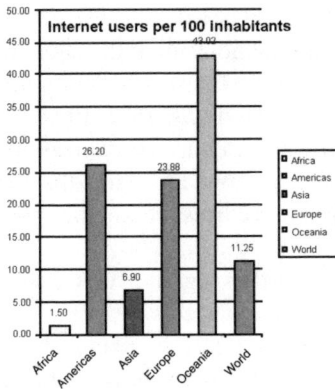

In the United States, 55 percent of the population used the Internet in 2003; in Europe, it was only 23 percent of the population, although the figure is growing rapidly (8 percent in 1999, 12 percent in 2000 and 18 percent in 2001). In both cases, it is likely that very few of those using the Internet were elderly.

The ITU also established that in Africa, there were only 8.65 telephones for every 100 inhabitants in 2003 – and yet to use the Internet you need to have not just a phone line but a computer as well!

So when we talk about communicating through the Internet, we are really talking about just 11 percent of the world population, 1.5 percent of the African population (although this last figure represented a 300 percent increase compared to 2000), 7 percent of the Asian population, 26 percent of the

American population (which for the ITU represents North and South America – this includes 55 percent of the USA's population and 51 percent of Canada's), 24 percent of the European population and 43 percent of the Oceanian population (which for ITU includes Australia 56 percent and New Zealand 52 percent).

In northern and southern countries there is a growing disparity between towns and cities. In urban settings there is often a choice of solutions for high-speed connection. In rural areas Internet access is still reliant on traditional telephone lines.

In Europe the problem tends to be more of a cultural issue. Some 85-year-olds will sit down at a computer and use it with determination and zeal; at the same time some people in their 50s say that the Internet is "not a tool of their generation". In that case, the divide is "cultural".

Those running many church organizations are from a generation that did not grow up using the new technologies. They are often the victims of an intergenerational digital divide.

After these mostly negative observations, it is worth recalling some positive aspects.

In industrialized countries, where every month sees an increase in the quality and coverage of the new communications technologies, the world of communications is expanding rapidly. In developing countries, many organizations – not least the United Nations Development Programme (UNDP) and the ITU – are working determinedly to give everybody access to communications technologies.

The World Summit on the Information Society (WSIS), phase one of which took place in Geneva in December 2003 (phase two is planned for Tunis in November 2005) was an opportunity for various stakeholders to develop a declaration of principles and a plan of action for promoting the development of new and just technologies. The WSIS was a forum for complex debate between representatives of civil society, governments and commercial companies.

The Internet is a practical means of communication for countries that are isolated, because it promotes democratic access. Many activists involved in the protection of minorities have become adept at quickly reaching a large number of people. The Internet makes enormous amounts of information available to all at very low cost. If, for example, a church develops a database of liturgical texts, that information can easily be made available to sister churches in Canada, Switzerland, Belgium or Africa without incurring any additional cost.

Another indication of growth is the rapid proliferation of cyber-cafés[2] in developing countries.

Lastly, there are now so many Christian organizations on the Internet, including in the countries of the South, that it is impossible to count them, much less to know all their addresses by heart! Christian search engines are more necessary than ever.

Without any doubt, it is in the United States that the Internet is most widespread, both generally and among churches. Reasons for this could include: (1) the Internet began in US universities; (2) the community-based nature of US society – people are brought together on the basis of social criteria, membership of a movement or pressure group, or even a minority demanding its rights. If you want "virtual" visitors to be interested in a website, they must feel as though they are in a familiar environment. Community spirit makes it possible to give virtual relationships substance. The Internet bridges distance – the Web can rebuild communities by uniting people despite their separation in terms of geography.

The ecumenical dimension of the Internet

Because of its universal nature, the Internet is a medium that favours the development of ecumenical relations. However, for many people it remains an undiscovered world. We are a long way from knowing and utilizing the full potential of the Internet, but the emergence of a whole host of discussions and initiatives is an encouraging indication of the opportunities to come.

It is not always easy for the Internet to find its way into our churches. The reasons for this are quite simple: most ministers and church or parish officials prefer face-to-face contact rather than long-distance contact, and their training in the new technologies varies considerably and is sometimes non-existent.

Something else that makes it difficult for churches (or ecumenical organizations) in Europe to engage in collaborative projects is that the people who administer such projects in churches or church organizations do not often meet; for the most part, it is ministers or church leaders who get together. They tend to concentrate on establishing good relations rather than on developing collaboration between churches on specific practical projects. As a result, different institutions develop similar information technology systems, when it would be logical for them to work together. Geographical distance, cultural differences and language difficulties contribute to this, as does the ignorance that comes from a lack of opportunities for organizations to exchange views on this subject.

Lastly, there has been the issue of "confessionalism": the desire to use the Internet to disseminate an image of the particular church, and the fear of losing control over a project shared with others.

However, technological collaboration can make for an exciting ecumenical project that does not necessarily lead to the problems normally linked with differences between denominations.

EXISTING WEBSITES OR PROJECTS

Most churches and religious organizations in the northern hemisphere now have an Internet site. Suddenly, websites that set out with the aim of communicating find themselves drowning in a sea of websites, each more attractive than the next and all trying to attract that same section of the population that currently logs on to the Web. One solution is to come together so as jointly to bring some order to this chaotic world.

An example: the World Council of Churches

The World Council of Churches (WCC) launched its website in 1995, in order to present the organization to members of the public interested in the WCC or in ecumenism in general *(www.wcc-coe.org)*. The site gives basic information on the ecumenical movement and the key signposts in its history. It features documents and statements on important issues the churches are dealing with, and publishes news and updates on current events in which the Council is involved. Most of the site is translated into German, French and Spanish, and some texts appear also in Russian.

With its growing number of pages (more than 7000 in 2004), the site is becoming complex and difficult to manage. The site-map is being rebuilt in order to give more visibility to the issues than to the structures of the organization. The new site will make considerable use of database technologies and content management systems. The idea is to make it a portal on ecumenism as seen by the WCC.

Apart from the main site, the WCC operates several smaller sites (described in appendix 2) in order to give more visibility to important ecumenical initiatives, such as the Decade to Overcome Violence (2001-10) and the Ecumenical Accompaniment Programme in Palestine and Israel (EAPPI).

The site *www.ecuspace.net* was created as an interactive platform for sharing information and ideas among ecumenical partners. Staff of ecumenical organizations may register and obtain access to shared forums. Most of the activity of the site, however, takes place in private forums which are used by specific groups to share their working documents.

The European Christian Internet Conference

In Europe, there is an outstanding example of ecumenical cooperation regarding the Web. Every year, the European Christian Internet Conference (www.ecic.info) brings together representatives of Christian organizations working in the Internet field, so that they can share their experiences – good and bad – and improve their understanding of recent developments and future potential.

Christian portals

For those who are on the look-out for something new, Christian portals are the point of entry. There are links to

Web sites in the United States and in Europe; the difficulty is in identifying the organization behind these sites, above all in terms of their religious affiliation (it is one thing not to know whether it is Reformed, Lutheran, Orthodox or Roman Catholic, but what a disappointment when a portal that looked like it might be interesting turns out to be run by a dubious sect!).

Another problem is the commercial attitude taken by some Christian sites that chase after Internet users like companies trying to corner a niche in the market.

Information about ecumenism

There are two ways of finding Internet information about ecumenism. First, many organizations (such as the World Council of Churches or Ecumenical News International *(www.eni.ch)* offer a great deal of ecumenical news. The website of the World Council of Churches includes links to several other ecumenical organizations. The WCC is linked to its member churches (many already have a website), to many other Christian bodies such as regional ecumenical organizations (present in Africa, Asia, the Caribbean, Europe, Latin America, Middle East, North America and the Pacific), national councils of churches, Christian world communions representing the main church families (Orthodox, Catholic, Lutheran, Reformed, Anglican, etc.), various international ecumenical bodies and church specialized ministries and ecumenical agencies. The WCC is also closely linked to sister organizations, such as the ecumenical humanitarian organization Action by Churches Together or ACT International *(www.act-intl.org)*; the Ecumenical Church Loan Fund *(www.eclof.org)*, specializing in micro-credits; or the Ecumenical Advocacy Alliance *(www.e-alliance.ch)* specializing in advocacy campaigns on HIV/AIDS and trade-related issues.

THE ECUMENICAL POTENTIAL

When you realize the potential of the Internet, you will realize that all the initiatives that we have given as examples so far make only limited use of the Net; they use the Internet's capacity for establishing links with other people – which is fantastic, but does not make full use of the potential for working together. In other words, rather than competing to produce a website that is more impressive than our neighbour's, we should be asking how we can use the Web to embark on joint projects. Those who appreciate this aspect cannot help but be enthusiastic about the prospects that the Web offers for the future. Imagine that a church wants to make a small database of liturgical texts: it might be useful to know that one already exists in a neighbouring church that shares the same language and the same theological approach.

Of course, the idea of cooperating with other churches is thinking outside the box, and might pose a problem to some people – but it is also a challenge laid down to those of us who preach a gospel without borders and who want to build a universal church.

Why not consider sharing liturgical texts with churches of the same language and of the same denomination in other countries? Why not consider a collaborative project of this kind between Protestants, Catholics and Orthodox within the same country?

These ideas might seem idealistic, but they show the extent to which the Internet challenges our convictions. We all, as Christians, have a dual responsibility in the face of these fascinating challenges.

First, we are currently seeing a real surge in initiatives of all kinds, but in particular of Christian directories, on-line Bibles, discussion forums; it is our responsibility to make sure that these initiatives are instilled with true Christian values – values that, in a world where the pursuit of pleasure, the acquisition of wealth and their cultural byproducts are largely accommodated, are based on generosity, openness, transparency and simplicity. *Let us instil the Internet with Christian values*.

28

Secondly, we have at our disposal an incredible tool for sharing. Ecumenical mutual assistance organizations, for example, that once came up with the idea of the sharing of finances – and subsequently of other resources – are now exploring possiblities for information sharing. *The information I have is not a means of asserting my power but is a responsibility that is entrusted to me so that I can make it available to those who need it, for the good of everyone.* Such organizations are currently working on a way of sharing their information via the Internet. Instead of trying to use the information we have to make ourselves richer, we should be giving it to sisters and brothers who will make good use of it. This is a big challenge, but it also offers hope to the partners in the South who are so remote that they might be able to overcome the exclusion imposed by their geographical isolation.

It is a challenge for all our churches, for they will have to learn to think about both aspects of information-sharing: what they can gain and what they can give.

In conclusion to this chapter, we note that *the Internet is not a fashionable new gadget, but a new space that we must occupy.* This is implied by the etymology of the word "ecumenism": "oikoumene", means "the whole inhabited earth", or "all the world" in the sense of Luke 2:1, "a decree went out from Caesar Augustus that the *oikoumene* should be taxed". We in the ecumenical movement are called to ministry in all the world, real and virtual.

NOTES

[1] The term "John Calvin" requests documents in which the word John is immediately followed by the word Calvin. If you ask for John Calvin without quotes, you get all documents that contain both words; for example, a document on Calvin Coolidge who did a post-graduate term at St. Johnsbury.
[2] Places where you can use a computer and access the Internet for the price of a local call.

2. The Internet in the Current Social Environment

The birth of a new culture

The Internet is certainly a very powerful and seductive tool, one we must learn to use, but it is much more than that. If it were just a matter of learning how to use a new technology, then the Internet would not provoke such extensive debate, or opposing views. There are a great many economic, political, sociological, anthropological and religious questions that are raised by everything relating to virtual reality, and to the new information and communication technologies.

Several characteristics make the Internet an important phenomenon: its power, its universality, its lack of physicality and its interactivity. Then there is the quantitative aspect: at some point, the omnipresence of the Internet means that we find ourselves facing more than just a medium – this quantitative change leads to a qualitative change, and the technological tool is transformed into something much more than that. Lastly, there is the fact that this new method of communication falls within a wider context of communication theory, and has spawned a new culture.

Philippe Quéau, an expert in the study of new images and new technologies and director of UNESCO's information society division, says that, to the extent that the technology is on the point of being mastered, the era of pioneers is over, but a vast field of activity and study is opening up that is cultural rather than technological in nature. This is what he calls the "social dimension" of information and communication technologies. The new Internet and virtual communication culture is a result of two converging phenomena: first, the mass availability of computer equipment – the explosion of digital technology, and secondly the low-cost computer terminals that make the Internet accessible to all. There has been a popularization of new technologies in recent years – and this phenomenon continues – along with a great deal of commercial advertising. But the new culture that is forming before our eyes is also a result of the emergence of new types of images, messages and languages: "A new image culture has emerged, which no longer stems from the televisual image but from the live, interactive, real time, networked image."

A new type of image is emerging, characterized by a relationship to reality that is no longer "analagous" but "simulatory". According to Quéau, "the image has become as written and as formal as words. It is now realized that the relationship that matters is not between paper and the screen, but between the formal and the informal, between the sensory and the conceptual." We have to learn how to read these new images, which is why Quéau talks about "new writing", a "new alphabet", and about the need for a "new literacy". We must learn how to "read", i.e. how to find our way around this "mental navigation within infinite information spaces". It really is very much like learning to read. It is an apprenticeship in two senses, because although we need to be familiar with the new technological and cultural environment in order to be able to use the new media properly, we also need the new media if we hope to have a proper understanding of the increasingly complex and prolific information that is communicated to us about the society in which we live. The image has become an intellectual tool for knowledge.

This technological, economic and linguistic revolution has important consequences. Quéau observes that we cannot simply watch this emerging new culture, we also have to "civilize" it. He is referring to the economic and social upheavals that resulted from the arrival of new technologies onto the mass market. He also raises other questions, like the need for a new way of learning to read and write. Following the era of the printing press (books and newspapers) and that of the image (photography and television), we are now entering a new "digital" age.

Another way to define this emerging culture created by the new communication technologies can be found in the writing of Manuel Castells, currently professor of sociology at Berkeley University. He argues that this new culture, centred around the notion of "hypertext", can be defined by the coexistence of at least five elements:

Integration: the combining of artistic forms and technology into a hybrid form of expression. *Interactivity:* the ability of the user to manipulate and affect her experience of media directly, and to communicate with others through media. *Hypermedia:* the linking of separate media elements to one another to create a trail of personal association. *Immersion:* the experience of entering into the simulation of a three-dimensional environment. *Narrativity:* aesthetic and formal strategies that derive from the above concepts, and which result in non-linear story forms and media presentation.

This new culture is so different from what came before that some people predict it will bring about profound changes to our society. It is not just a new culture or even a "revolution" that we should be talking about, but a new era. "The Net is much more than a new technology, it is a civilization shift," said one head of multimedia for a major international bank. In comparison with the introduction of rail, electricity or the automobile, the Web is a major innovation. The Internet is more powerful and more universal than any of these other inventions, and makes it possible (hypothetically – not in all parts of impoverished Africa, Asia, etc.) to be in virtual contact with any person or any reality irrespective of distance, to the point that distance becomes irrelevant. The universality of the Internet can also be understood in a more human sense, in that the Internet affects every sphere of activity – work, family, leisure, research, consumption and so on.

This civilization shift can be seen in the way the Internet changes how we behave both individually and collectively.

Several people have tried to come up with a more comprehensive definition of the new culture that is emerging from the "industries of the third millennium". Joël de Rosnay, executive president of Biotics International, defines it as both "fractual" and "hypertextual":

> Fractual because each of us, according to the "density" of our culture, will build a micro-organism of the whole. Hypertextual, because this form of culture will be connectional, linked to the infinite to a different degree from the fractual culture. This form of culture is both personalized and general, individual and collective, respecting each person's diversity and temporal (cultural) density, which is the basis of relations between the symbiotic human being and the macro-organism of society.

This definition prompts us to look at one of the big innovations of the Internet, the emergence of a seductive imaginative world, combining scientific truths, science fiction, fantasy and futurology. Pierre Lévy, a Canadian sociologist at the University of Quebec Trois-Rivières, goes as far as to say that this new culture will change not just humankind, but also the most mysterious, secret and unattainable aspects of life itself:

> The emergence of cyberculture is profoundly changing the way in which the world of form exists, changes and is transmitted to the human mind. Relations between minds are changing. The nature of matter is changing. Humanity is mutating.

Humankind is changing in response to the new information and communications technologies, changing mentally and psychologically – so why not physically as well? We have moved from description to fiction, from science to science fiction. The new media are capable of mediation, even universal mediation.

The ideals surrounding the Internet

Some people try to predict the world of tomorrow – describing the new civilization being woven by the new information communication technologies as "cyberspace", "cyberculture", "noosphere",[1] or "symbiosis". Are the people who use these futuristic terms part of the avant-garde? Are they people with exceptional foresight? Or have they been taken in by the fantasy of the ultimate power they tell us about? Reading all that has been written – often in scientific or pseudo-scientific jargon that is difficult to understand – one would tend to think the latter.

Either way, it is significant that two diametrically opposed attitudes are crystallizing around the Internet: on the one hand, a utopian, fantastical, enthusiastic vision of a liberated and liberating future; and on the other hand, strong condemnations, caveats, categorical denials of an ideology that is seen as totalitarian and surreptitiously advancing through the new technologies.

Joël de Rosnay is typical of the "utopians"; as early as 1995 he talked about the Internet as being "the genesis of a global brain". Elsewhere, he writes,

> The hybridization of technologies is a factor in the acceleration of the co-evolution of humankind and the machines that process information. With the growth of multimedia, the electronic highway, interactive television, interpersonal networks of global computerized communication, we are witnessing the construction of the global brain and nervous system of the societal macro-organism.

What is remarkable in this kind of discourse is the hypothesis of the disappearance of boundaries between humans and machines: humans becoming mechanical and machines becoming more human. Also, there is the idea of "de-realization", the blurring of the borders between the real and the virtual so that we are no longer very sure which world we are in. There is often substitution, with the virtual becoming the true reality. Thus, de Rosnay talks about a "real virtuality", a concept that for him means a kind of de-realization, a way out of the limits of our own bodies: "Freed from the material constraints of our bodies, we explore at our leisure unknown worlds, inhabited by virtual clones and multifunctional software robots – while evolving in dream landscapes." We are in the land of science fiction.

Conversely, there is the development of a global, holistic, total if not totalitarian way of thinking. We can no longer escape the larger Whole. Typical of this type of thinking is Pierre Lévy's recent work. Lévy is an apostle of the new world created by information and communication technologies:

> For the first time, the Web is operating on the scale of the human race, a potential mediator between all human beings. The wide fabric of meaning is materializing before our eyes... The Web heralds and is progressively realizing the unification of all texts into a single hypertext, the fusion of all authors into a single, collective, multiple and contradictory author. There is now only one text, the human text.

Though the author never defines what he means by this term, globalization is omnipresent. There is no longer room for individuality, social particularities nor the unexpected incidents and hazards of a personal or collective story.

It would not be surprising if this discourse tended towards religiosity. It may be vague and universal religiosity in which everybody can find him- or herself, in such a way that there is often a deification of the human, with human beings acquiring powers and knowledge that until now have only been attributed to God: omnipresence and omniscience; crossing over the boundaries of time and space; immateriality, and so on. Lévy states this explicitly: "Technology is a foundation for the spiritualization of the human." Lévy first of all identifies three "anthropological objects" which, since the dawn of time, have contributed to the humanization of the human being: fire, art and writing – to which he adds a fourth, more recent one: the computer (which he for some reason also calls cyberspace). But of these "anthropological objects", cyberspace (this poorly defined term in the end means "the world of the Internet") attains a different level in the universalization of humanity and of consciousness. Lévy attributes qualities to cyberspace that are in fact those of the divine: qualities or characteristics that are granted only to God. "Cyberspace is a sort of objectivation or simulation of the global human consciousness... cyberspace is the vast, dynamic reserve of all forms of interaction, the form of forms, the idea of ideas." Cyberspace, which Lévy sees as if it had a divine nature, brings together the entirety of human endeavours, which Lévy summarizes in three words: language, technology and religion. Religion is thus once more a feature.

From the technological point of view as much as from the human point of view, we are witnessing in Lévy's work a process of divinization. Lévy's holistic thinking brings us surreptitiously from observed reality to imagined reality, from thinking to believing, from the humanization to the divinization of humankind. This is a good example of the suggested close relationship between the Internet and religiosity. It is also this that is denounced by critics of the Internet – or to be more precise, critics of the ideologies surrounding the Internet.

However, the utopians should not be dismissed as being manipulative or as frauds. Patrice Flichy, researcher at the CNRS, in a well-researched book called *L'imaginaire d'Internet* [The Internet's imaginative world], shows that there is a necessary and positive utopia that is the driving force behind all technological invention. A utopia is a positive and creative value, and it is utopias that motivate the scientific community. Therefore it is not surprising that the Internet should, among other things, spawn new utopias, when we are talking about the emergence on a mass scale of an entirely new, previously unknown reality. The fact that the Internet makes people dream proves its reality and its power.

Triviality and omnipotency of the Internet

Other theorists are far from sharing Lévy's enthusiasm and warn against the various spin-off effects produced by the invention and popularization of the Internet. The criticisms made by two writers in particular – Dominique Wolton and Philippe Breton – seem to complement one another. They have both studied and researched communication and the media for some time.[2] Wolton criticizes the "technological ideology" that accompanies the Internet phenomenon, while Breton, also of the CNRS and the Sorbonne, dislikes "the messianic ideology" that it brings. These are issues to which the church should not remain indifferent.

There is no prospect of a society being created around the Internet, says Wolton: invasive advertising has obliterated its social calling. No real thought is being given to this new multimedia tool. Business has ousted humanity. Commercialism has taken over the Internet. Thus we can talk about an ideology of the Internet, but it is a technological or commercial ideology. Wolton criticizes this technological ideology and portrays it as being based on three hypotheses:

1. Technological change is synonymous with progress, above all in the communications field.

2. We have to change now, or it will be too late.
3. Criticism of any kind is synonymous with fear of change and defending things that are outdated.

However, Wolton does not merely condemn the ideology that accompanies the Net, he condemns the Internet itself by refusing to recognize it as a medium. He says the Web is not part of the media because "the existence of media always goes back to the existence of a community, of a vision of relations between the individual and collective and to a certain representation of the audiences". Unlike radio or television, the Web has no strategy for communicating with an audience; it can therefore be conceived of only in relation to its technological capacity for transmission. Moreover, the Internet is on the side of business more than on the side of democracy, and this makes it a threat to democracy, for the consumer takes precedence over the citizen.

Other experts in the traditional media share these reservations. Emmanuel Hoog, president of the French National Audiovisual Institute, is worried about the extreme fluidity of the virtual network. Does the Internet contribute to wiping out memory? According to Hoog, the seven characteristics of the Internet (immediacy; volatility; technological instability; immensity; mixedness; hypertextuality; non-finitude of objects) are all problematic from a point of view of the management and storage of information, in that they all seriously impair the establishment of a digital memory. That memory is made difficult if not impossible to store due to the ease with which one can delete the traces, make references disappear, manipulate the evidence, privatize the collective memory. However, there ought to be a memory of this "media revolution", of which the Web is a central element.

Wolton points to three worrying phenomena created or amplified by the Internet to which the church, in developing a strategy of embracing the Internet, should try to respond:

• *Interactive solitude:* The more technological communication there is, the less human communication there is. "In a society where individuals are freed of all rules and constraints, the ordeal of solitude is real; similarly, it is painful to become aware of how difficult it is to make contact with others." We will come back to this move from individualism towards solitude in relation to the Internet, later on.

- *Time is short-circuited:* Any act of communication pre-supposes the passing of a period of time, but because the new media work in real time they have a tendency to erase that time-frame.
- *The impossibility of transparency:* It is an illusion to believe that one communicates better and more easily via a screen, because social interaction is never transparent.

We have to give thought to these issues. Wolton concludes,

> The idealization of technology, which can be seen on a daily basis, is a result of the weakness of our theoretical culture on communication issues, and more generally of our society's fascination with technology.

He says we must think about communication, for communication is major reality of the 21st century, but simply stating that does not of itself allow us to think about it in depth.

We have to know how to think about communication, including its political, social, economic and religious aspects, and about how to integrate it into an overall vision of society. In addition to the lack of analytical reflection about the new media, a lack of religious education means that there is a tendency to confuse the immateriality of belief with the immateriality of the Internet.

For many people, these two forms of immateriality acquire adjacent, if not similar, spiritualities. Faith and new technologies are both based on the immaterial. But is it the same kind of immateriality? Believers will, of course, say that it is not, but for those who do not know much about religious questions, there is more than just an analogy, it seems to be two different forms of the same phenomenon. The first comes through the medium of human beings, the other through the medium of technology: human beings develop a spirituality that seems to be found, in an amplified form – a

purer and more original form, although a more recent one –
in the new technologies. All that is immaterial has a spiritual
connotation, which means that it is difficult in the end to be
sure of what is not "spiritual".

The "religious" nature of the Internet was one of the
charges levelled against it in an important book by Philippe
Breton, *Le culte de l'internet* [The cult of the Internet]. Breton believes that the religious dimension of this medium can
be seen above all in the ideological, commercial and technological discussions that surround the promotion of the Internet. This is what he calls "the all Internet strategy", which he
says brings with it this new spirit of religions. The advertising debate is a good example. Countless adverts for computers, software, computer programmes or telecommunications
contain a manifest or latent religiosity – symbols, images or
objects, expressions, lighting effects, myths and experiences
evoke the presence and power of the sacred.

Breton does not reflect on the intrinsic nature of the Internet, although he tends to be distrustful of it. It is possible to
refuse to entertain these new beliefs and to use the Internet
rationally, but this minimalist use is marginal in contrast to a
more emotional use, supported by commerce and advertising. It is this second use that is apparent in the current spiritual practices of some sect-like groups or "New Age" religions. Internet users have similar attitudes, dreams and
mental constructs to those of new religions, says Breton:
detachment from the body, devaluing the material in favour
of the immaterial, holistic thinking, disdain for reason, ideals
of transparency, the application of the metaphor of light, the
quest for ecstasy, the search for universal harmony between
human beings, and so on.

Breton's views about the religiosity of the Internet are
clear-cut. This religiosity is to be viewed negatively when it
goes against the values of humanism. It is at one and the
same time omnipresent, dangerous and heathen. The figure
of the devil is recurrent: "The imaginary world of the Internet is full of dark forces that, simply because they are
obscure, and often only because of that, are, irrevocably, seen

to be on the side of the devil." And Breton concludes, "This religiosity appears as non-deist, spiritualist, dualist and anti-human."

These new "religious" themes are also diametrically opposed to a Christian way of thinking, because the values of Christianity and humanism go together, the latter having its roots in the former. If this religiosity does exist, it is a religiosity that has nothing to do with the major principles of Christian theology: "This new religiosity is, if not atheistic, at least indifferent to the idea of God." It is a spirituality without God that rests upon the deification of human qualities, or even the deification of technology or matter and upon the immaterial flow that is too readily described as spiritual influences.

The acclaim with which some Web users and theorists have hailed the thinking of the French Jesuit, Teilhard de Chardin, brings yet more confusion to this Internet religiosity, which is as disorderly as it is widespread. Many are fascinated by the great cosmic synthesis suggested by Teilhard de Chardin. He was the first to propose a synthesis of physical and biological phenomena, long before the new technologies appeared, claiming that these phenomena were part of an evolution of the universe that would culminate in union and fusion with God. Teilhard is supposed to have predicted the emergence of the Net half a century before it happened, by talking about the rapid development of a "thinking membrane covering the planet" and unifying the human consciousness (the "noosphere").

This definition does correspond to a non-scientific definition of the Internet. John Perry Barlow, one of the pioneers of the Internet, adds to Teilhard de Chardin's definition of the "Omega point": "The idea of connecting every mind to every other mind in full-duplex broadband is one which, for a hippie mystic like me, has clear theological implications." Elsewhere he says,

> Back in the 1930s, Teilhard de Chardin wrote about the great teleological process of mankind, the phenomenon of humanity. And he said that the whole point was to create a global collec-

tive organism of consciousness that would be sufficiently smart to keep God company. I think we're actually headed there.

How better to express the ambition hidden by this "religious" ideal – through tireless technological invention, humankind might eventually become God's equal. Clearly, although Teilhard de Chardin is being quoted, his thinking has not been understood and is not respected. Instead it is used in such a way that it is turned into a set of references that were not those of Teilhard, for whom the "noosphere" was clearly something associated with God, and specifically with a Christian God.

We therefore arrive at intersecting viewpoints: strictly secular humanist ways of thinking about these new media would seem to be more Christian than religious or spiritual ways of thinking about them. Sectarian and "New Age" ideologies have more in common with each other than with traditional Christian faith or with a philosophy based on monotheism and the Bible.

That is not all. Breton also condemns an anti-humanist and anti-Christian ideology, not just in the debates that accompany the spread and the commercialization of the Internet but also in some of the characteristics that are fostered by the very use of the Internet. He says that it is the Internet itself, and not the debates about it, that promotes a vision of humankind and of society that is wrongly described as religious. He sums up this perversion of fundamental and universal values by pointing to their threefold covert and overt rejection by the new information and communication technologies. Those rejections clash head on with both an understanding of civil society and Christian self-understanding.

The rejection of the law: The demand for total and unbridled freedom clashes with our concept of the law as guarantor of the rights and responsibilities of all, with the aim of greater freedom for all. Moreover, the law plays an important role in structuring biblical thought. The law is a gift from God to a humanity confronted with prohibitions and boundaries, in order to make humanity more responsible.

The rejection of mediation: The illusion of transparency. Institutions are suspected of restricting the freedom of every individual. Any political system is suspect; we demand direct democracy, without any intermediary. Religiously, it is good to recall that Christianity feeds on mediation, whether it is by institutions (the church), professionals (ministers) or symbols (liturgy and sacraments).

The rejection of the embodiment of the word: The Internet is a realm of the spirit, the triumph of the soul over the body. Here too, it is worth noting that in the Christian system every word is embodied, whether it is that of humanity (human body, or the body of society) or of God (Christ, the human face of God).

To return to the religiosity of the Internet. Some Internet experts in the United States make observations that are grist for Philippe Breton's mill. In a brief historical overview of the first users of the Internet in the United States, it is noted that one of these communities of users and of researchers at the origin of the Internet entertained quasi-religious ideas, close to the "New Age". Breton describes the state of mind of some Internet luminaries as "megalomaniac techno-spiritualism". They sometimes went as far as to claim, "We are like gods, we will act accordingly." Jörg Hermann and Andreas Mertin, two German theologians, quote the Czech philosopher Vilém Flusser, who goes even further in the deification of humankind: Internet scientists have succeeded where God could not, because whereas God partly failed in the creation of the world (the world is imperfect and characterized by evil), scientists have invented a perfect technology, a flawless creation. Humankind has thus well and truly, and definitively, replaced God!

Some of the people who helped to invent the Internet have, from the outset, imagined that it might help to propagate ideas of a new religious utopia. Historically, the Internet was therefore not just a tool at the disposal of the scientific community, but was also the expression or the vehicle of a new religious utopia – based on a religiosity that is diametrically opposed to the thinking and spirituality of Christianity.

Breton's criticism of the religiosity connected to the Internet is therefore most pertinent, but is he justified in identifying religiosity and interactivity, divinity and virtuality at first glance? Isn't he exaggerating the importance of a phenomenon that is without doubt very real (the emergence of this new religiosity) but which in itself may not be the key to an understanding of the new media? Doesn't the fact that he distinguishes between two different and opposing uses of the new media (a "religious" use, the All Internet; and a humanist use, the "enlightened" use) go against the idea that they are *inherently* religious? Are the links between the new religious movements and the new media as clear as all that? After all, everything that Breton says about the emergence of this "spirituality without God" in the context of information and communication technologies could just as easily be said about other sectors of life: advertising, political relationships, parallel cultures, artistic expression, sport ("football as religion"), etc. Information and communication technologies are not the cause of this religiosity but are, like other aspects of our society, just the victims of the concomitant emergence of an irrational element in our world.

Breton is right to alert us to the fact that these new media are not a neutral object or technology. Media have a number of effects not just on the user but on society as a whole, as a result of changes they entail in our way of thinking and our way of life. He is also right to emphasize that, among the many effects brought about by these new media, there is much that is irrational or religious, a mish-mash of elements borrowed from various forms of spirituality. There is an undeniable need for theological clarification.

NOTES

[1] The word "noosphere" was coined in analogy with the "geosphere", the world or layer of lifeless matter, and the "biosphere", the world or layer of living matter. Beyond and superimposed on these spheres lies another dimensional sphere, the "noosphere", from Greek "noos, nous" = "mind", and "sphaira" = "globe", a figurative envelope of con-

ceptual thought, or reflective impulses produced by the human intellect. It is not scientifically measurable, of course, but its presence is
strongly felt and its influence is all-pervading.
2 This is not the case for all critical approaches. Alain Finkielkraut's
attack in *Internet, l'inquiétante extase*, is flawed by the fact that the
theorist is obviously not sufficiently familiar with the culture of the
new technologies.

3. How the Church Assesses the Internet

What position should the church take towards the new media? How should it evaluate them? It seems to us that there are two pitfalls the church must avoid:

1. *Over-hasty enthusiasm:* showing a naïve enthusiasm on the grounds of usefulness or keeping up with the times.
2. *Undifferentiated rejection:* rejecting these new technologies completely and without differentiation. Should it do this, the church would risk social marginalization, losing contact for ever with those whom it is already having difficulty reaching: young people, those on the margins of the churches, "non-practising believers".

The church must therefore think about information and communication technologies in the same way that it has to think more generally about communication. Theology has to find the link between the foundations of faith and current practice, between the tradition and the current situation. Roland Kauffmann, a theologian interested in building a relationship between theology and information and communication technologies, says,

> The new communication technologies represent a challenge for theology, in so far as computers have for the first time crossed the boundary between being simple automated tools or means for real social, political, cultural and economic communication. The influence of this means of communication on our way of thinking, our understanding of the world and of relations between individuals must be the subject of serious theological analysis if theology is not to become completely obsolete in a computerized society.

Constructing a critical theory about the Internet is all the more important given that it is a medium that tends to minimize time, and therefore thought. "Internet users risk limiting themselves to spontaneous reaction rather than giving themselves time to think," observes Dominique Wolton. Wolton makes an explicit appeal to churches to give the new technologies a human face. He thinks that the church is being too cautious, too reticent in the face of the "communication revolution". The church needs to be critical of the technological ideology of communication for the sake of sustaining relationships on a human scale. Therefore, the church should recall that a normative vision (or ideal) of communication exists, and not just a functional vision (or technology):

> If the church is to play a role in the globalization of communication it must of course learn how to use the new media, but it must above all remind people that it has the capacity to set out a human vision of communication that is more important than merely a high quality of technological transmission.

To put this in concrete terms, we have picked out five things linked with the Internet that might be in conflict with the message of the gospel and the church, as well as with any humanist, non-instrumentalist vision of human beings.[1]

The means becomes the end

With the Internet, more than with any other medium, there is a risk of the medium itself becoming the message. More than ever, the words of Marshall McLuhan risk coming true: "The medium is the message."

This Canadian academic who converted to Catholicism and who died in 1980 is now considered by many people in the multimedia field to be a modern prophet. He was the first person to think about the culture based on the mass media. McLuhan was thinking about the new media of his time, which were television and video, but the new media of today, far from rendering his thesis obsolete, strengthen it. First, he spelled out the link between the technological revolution and changes in civilization. He discovered that any technological medium and any mass media not only influenced thinking, but changed it, kneading and shaping it. This is where the title of his last book, *The Medium Is the Massage*, came from. The media are responsible for a new form of thinking (and not just of doing things), which is itself responsible for new human and social interaction:

> The medium or process of our time – electronic technology – is reshaping and restructuring patterns of social interdependence and every aspect of our personal life. It is forcing us to reconsider and re-evaluate practically every thought, every action and

every institution formerly taken for granted. Everything is changing – you, your family, your neighbourhood, your education, your job, your government, your relation to "others". And they're changing dramatically.

Like the conventional media, but even more so, the new media constitute a message in themselves, and there are three reasons for this:

1. The Internet is a very powerful and very seductive tool: we are therefore all the more willing to decide that it is fun.

2. We are no longer within the conventional model of communication based on transmitter, receiver and message. Everybody is communicating with everybody else: the message disappears under the duplication of transmitters and receivers.

3. The message itself is multifaceted: there is no longer *a* message, but an infinite number of messages – as many messages as users, because each user produces his or her own signals; suddenly, there is a risk of there not being any message at all. The means become the end in itself.

McLuhan's prophesies have often proved to be true, but that does not make them any less problematic, nor any less in conflict with a Christian vision of humankind and of society. Christianity claims to have a message, a fundamental message, that goes beyond any kind of medium (and even beyond human beings themselves as a medium that produces messages, both verbal and non-verbal) in so far as this message is based on God and on God's word, and not on human achievements. This direct opposition between McLuhan's theory and the objectives of Christianity is very nicely summed up by Claude Demissy:

> The bond created by the Internet will not be able to be the mysterious symbol of the communion of saints in the way that bread, wine or water are able to symbolize the activity of the Divine... It is the way in which God is present in the midst of human beings that is the object of the mystery and not the virtual link between men and women and the church community.

Despite McLuhan's professed Catholicism, we find it easier to support Dominique Wolton's argument when he says that "the Internet project is too often reduced to technical equipment! The result?... The human dimension always seems less important than the technical performance or the marketability." This media sociologist, who defends a humanist vision of communication, is right to remind us that technology cannot stand alone: "It is the message that is important, and not the way in which it is delivered... The medium is a project, not a tube."

Leaving aside these theoretical considerations and taking a more practical approach, the means replaces the end when the only objective of the Internet becomes the Internet itself. We know that the Internet can become a drug, or something like it: "cyberdependency" is said to affect 6 percent of Internet users. In an article about how the Internet can serve the churches, François Diot proposes, by way of a safety guard, that Internet users should always ask themselves three questions, the answers to which will help to avoid becoming "cyber dependent":

1. Am I at any moment able to turn off the computer and leave a world that continues to live without me?
2. Am I able to end the contacts woven through the Internet without feeling abandoned?
3. At what point do I consider that I am losing more in relationships and in knowledge than I am gaining in the field of communication and information?

Reality makes way for virtual reality
Nor is this phenomenon new: it existed in the already antiquated formula of the "entertainment society" or the "public image", where the representation of a reality tended to become more real than the reality itself. First French filmmaker Guy Debord, then Jean Baudrillard, sociologist and philosopher, condemned this phenomenon which emerged alongside the mass media. But the new technologies amplify this, making everyday reality less true than the virtual; giving virtual reality first place and letting it become the basic ref-

erence for relationships with others, and with reality itself. As mentioned earlier, Philippe Breton alerted us to the risk of physical separation, the end of face-to-face meetings. Little by little, we see the idea emerging that it is possible – and that it would be a considerable step forward – gradually to leave one's body, or even to give it up entirely, in favour of a virtual body. Human beings become purely souls. Pierre Lévy talks about "the interconnection of consciences"; others talk about the possibility of "cyber-relationships", even "cybersexuality".

Patrice Flichy develops this idea in a more nuanced way. Nuanced, because he does not systematically oppose reality and virtuality. The virtual is not viewed in the same way by everyone. For some people it is a devaluation of reality, whereas for others it represents a fulfilment of reality – another, more perfect, form of reality.

Flichy sees two types of virtuality. First, a *positive* virtuality that feeds on an ideal, and is therefore stimulating; this virtuality is a way of giving greater depth to reality, daring to create new things on the basis of *simulations*; it is at the root of many scientific experiments or artistic creations; making it possible to "simulate situations", to "create new worlds", to "escape reality to create imaginary worlds". This is a vital kind of virtuality. Virtual reality does offer people the information technology tools to help them dream. Information technology is not just a set of machines, numbers and functions, it is human interaction with the technology that will make further innovation and creativity possible. Flichy points to three positive values of the virtual associated with virtual reality: simulation, interaction, immersion.

However, there is also a *negative* virtuality, which can be compared to an ideology (rather than to an ideal), in so far as it offers an escape from a reality that is seen as inadequate or not worth living. It is essentially this side of virtual reality that we should be wary of, because it aspires to replace, in its entirety, life itself.

Flichy's analysis of virtual reality is well documented. He presents the groups and initiatives that are behind the development of the concept: the "hackers"[2] – the computer fanatics who live only for and via these new networks and neglect all the rest; the science fiction writers; the MUDs (Multi-User Dungeons), in which the participants build an imaginary world together; and lastly, the "cyborgs" – a word created by two scientists out of the words "cybernet" and "organism" to describe the design of an individual capable of personal evolution within an artificial environment.

Without going into the complexities of the dual form of virtual reality (true virtual reality and imaginary virtual reality), we will focus on four spin-offs that are in direct conflict with the Christian faith, which sees itself as being in touch with the realities of the world. In more ways than one, the Christian response to these spin-offs begins to establish a "theology of reality".

The self is multiplied

To begin with, there is the idea that virtuality comprises "a series of representations that can be multiplied ad infinitum from a self that is nonetheless unique". One and the same person develops multiple sensations and explores different worlds simultaneously. From there, it is a short journey to the idea of the "multiplication or replication of the body", as a result of the fact that, thanks to the Web, we can be in several places at the same time, in real time. Not only can we change identities and define ourselves as a new fictitious person, but we can also leave our own body and become somebody else. The idea of reincarnation is not far off. We can also acquire several identities simultaneously, and be different people at the same time. We can even become human and non-human simultaneously, because some programmes develop an imaginary world where some of the words are not produced by human participants, but by machines.

Reality is devalued or rejected

Many programmes are created on the basis of the idea that "reality is no longer enough". It is therefore about transporting us to another, richer and more complete world – the world of

50

cyberspace. These new virtual worlds replace the old, devalued, real world. Virtual reality is no longer an extension of reality; it is replacing it. The conventional media and the new technological media are therefore completely different in nature, "newspapers and radio report things, theatre and cinema show things, cyberspace takes you there". It is a matter of taking us somewhere else – into another life, a new life. There is something better than life, and that is cyberlife, where the shocks, bumps, arid stretches and failures of real life have been eliminated. In cyberlife, everything is sleek, easy, smooth and available. Nothing stands in our way any more. We can "live on our computer screen, take action through the keyboard and the mouse, without taking the risks of real life".

The body is devalued

One of the effects of the emergence of cyber-reality is the devaluation – the instrumentalization even – of the physical body of the Internet user. At best, the body is a nuisance; at worst, it is an enemy. We return to a dualist vision of human beings, based on spirit, thought and idea, and not a body with five senses. We encounter a world-view close to gnosticism, which is distrustful of the body, the human being, the weight of reality. We can therefore easily leave behind our own bodies, be outside spectators of our own emotions, watch ourselves living. One of the effects of this is the devaluation of human feelings, beginning with sexuality, which can be experienced in a totally mechanical, artificial way through the intermediary of the machine. This is "cybersexuality" which does not require any commitment of a partner in the relationship, because the partner is primarily the sensations, emotions and tactile impressions produced through the computer, which we have access to by means of receptors. There is also an idea that cybersex is risk-free, because it no longer involves interpersonal contact.

The machine and the human merge

We have just seen an example of this, in the idea that the machine can serve as a mediator of tactile sensations, emotions and the expression of feelings. Our PCs could soon be directly connected to our brains. A distinction is no longer made between the mind and the machine, between the human and the technological: the two merge, sometimes becoming interchangeable; we talk about "downloading our mind". A whiff of religiosity is then grafted onto this idea, as, for example, the idea that we might be able to live as immortals, because we will have downloaded the contents of our brains – the intellect and the emotions – onto a computer programme. That programme could be saved, copied and stored, so that it could then be re-used. Thanks to several back-up copies, we could then be witness to the "miracle of electronic resurrection".

These four spin-offs of virtual reality are mostly fantasy. It is not virtual reality itself that causes problems, but its accompanying, ideological world-view. As Flichy summarizes,

> In the ideological camp, we find the idea that the individual can abandon his or her body, become a pure computer soul; that there is no longer any difference between the simulated and the real; and that we can surf through human relationships without engaging in them. Cybersexuality is the first example of this ideology... The imaginary world of virtual reality becomes an ideology when it imagines that we can put our bodies in parenthesis, that information technology can create a-corporal human relationships, unconnected from any engagement of the body in space and time.

In talking about virtual reality, it is important to distinguish between the "ideal of liberation" and the "ideology of illusion". Virtual reality is not always in conflict with reality; sometimes it can be a useful extension of reality, and open up that reality to new possibilities. At other times, one experience of virtuality is in conflict with another form of virtual reality.

Hypertext destroys text

Some people see hypertext as one of the most important aspects of the "Internet revolution". According to Castells, "the most innovative vision of the cultural changes in the information era is perhaps the one built up around the concept of hypertext, and the promise of multimedia". But what is hypertext? Here too, we are faced with a floating concept, an idea that does not mean the same thing to all those who use the term. We have found at least three different ways of talking about hypertext. We will start with its main use, before considering the technological and philosophical spin-offs.

1. Historically, the notion of hypertext was born in the environment of the Apple Corporation's ground-breaking Macintosh computer of the 1980s. The term was invented to describe a text in which some parts, when clicked on with a mouse, connected to other texts. Originally, this meant texts stored on the same computer, but the concept found its true application with the development of the Web, which made it possible to interconnect texts via the Internet that were stored across the four corners of the earth. It is this meaning to which Dominique Wolton refers when he defines hypertext as "the links" that make it possible to move from one text to another, from one website to another, and from one virtual world to another. For Wolton, hypertext is

> a text on an information technology medium, comprising words, some of which may be linked to other texts and other documents, of which the structure is not literary. The links are indicated on the screen by colours, underlining, marks or images, that make it possible to move from one text to another.

This idea has since evolved, and has become blurred because of the invasion of the text and its hypertext by other elements.

2. Hypertext was then used to refer to this mixture of words, images, sounds and signs that make the multimedia language a script that is as much visual as verbal, a complete language. Very quickly, we began to click not just on letters, but also on images, sounds, video clips. Hypertext has become this multi-format script, which combines different kinds of signs: visual, verbal or audible. But there is a paradox: the image itself becomes a form of writing, the image

becomes a new script. The reverse is also true: writing becomes more visual, becomes an image. The appearance of various fonts, playing with the plastic nature of letters, turns writing into an image. Many attentive observers have noted this overlapping evolution; writing becomes more visual and images become more textual. François Diot notes that "a new written language is emerging that makes use of the art of ellipsis, abbreviation and the symbol" and Philippe Quéau that "the image has become as written and formal as words". The image becomes writing, while writing becomes more visual. We have a less and less analogical relationship with the image, and a more and more iconic relationship with language. Just at the formal level, hypertext seems more attractive than text.

3. These multiple developments of a text that is more than the sum of its parts have prompted many philosophical reflections about the change that this brings. A third understanding is born, which is the opportunity for the Internet user to personalize his or her reading by creating his or her own multimedia text, by constructing it as he or she understands it, in relation to his or her needs or desires, or simply by clicking at random. Castells describes this phenomenon: "not multimedia, but the Internet-based inter-operability of accessing and recombining all kinds of text, images, sounds, silences and blanks, including the entire realm of symbolic expression enclosed in the multimedia system". And he goes on: "The hypertext is not produced by the multimedia... it is instead produced by us, by using the Internet to absorb the cultural expression in the multimedia world and beyond." Following this last approach, hypertext would be the notion of a personal text that is in the process of being constructed, by the intervention not just of the person who produced the text but also of the person who reads it. The roles are reversed, or at least interchangeable. The reader becomes writer, the listener or reader becomes actor. We can understand the success of multimedia, in allowing readers to construct their own text as they read. There is no longer a static text, with the meaning already bestowed at the outset; there

is a multiplicity of meaning that is constantly evolving and adapting.

But will hypertext destroy the text? Does text risk appearing rigid, boring and tired? Certainly – this can be seen in the fact that in recent years the conventional text is often given in the form of a hypertext in presentations and publications: the text is made pretty and disguised in a multimedia form. A conventional page might be composed in the form of a window appearing on the computer, the chapters become menus, the important words hypertext links, etc.

Some Internet theorists are noticing this change, and claim that the classic literary written text is now outmoded, to be replaced by a multimedia text that is, at the same time, a hypertext. Thus, for Lévy the concept of hypertext is the most important cultural change that is occurring in the technological revolution that we are experiencing. Lévy, who is one of the most enthusiastic supporters of the Internet, goes so far as to claim that the riches of the Web surpass those of writing. He says that writing is the universal plus "totality", whereas the web is universality minus totality. For him, totality is a negative concept, a sort of locking up of the meaning by institutions, whereas today the individual aspires to rebuild "partial totalities" – the individual both wants to help to build cyberculture and to suggest its meaning. The Web is a personalized form of writing, because with the Web, communication is interactive and reciprocal. With the Internet, no message is any longer out of context. The context is immediately at hand through links to any part of the Web. Everything is contextualized and or can be contextualized. The multimedia is thus a superior, more complete and more human language than writing is. For Lévy, "the universal no longer totalizes [or restrictively defines] meaning, as is the case with static writing; it connects by contact, by general interaction, it tests itself by immersion".

Unlike Lévy, humanists such as Patrick Soriano or Alain Finkielkraut are worried by this development, and point to the hypertext as the source of many dangers if it claims to replace the classical text, in that its logic interrupts any lasting form of linear reading.

We can see the difficulties with the notion of hypertext for biblical Christianity, whose faith rests essentially on belief in scripture and a tradition that wants to use the meaning of that scripture in history. The Bible is certainly a collection of extremely varied texts, but it is not a hypertext, despite what Lévy claims. As a printed book, there is very little interactivity. However, the cross-references to comparable passages, added by editors, do give the impression of a text that is alive, made up of reminiscences and allusions, the many traces of the Bible's roots in a highly interactive oral culture. Today this is expressed in the meaning that is constructed in the act of receiving the message which is read, interpreted and symbolized in the liturgy and during worship. There is also the fact that the austere, classical, uniquely literary form of the Bible risks becoming more and more of an obstacle to accessing the root and content of its message. We see two ways of responding to this problem, one material and the other spiritual:

- Materially, it is completely possible to translate and transcribe the Bible into an interactive form. We are only at the beginning, but on-line Bibles have without doubt a bright future ahead of them, in line with the development of books.[3] Because today we get to the content via the form, we have the opportunity to make the form more attractive, after which we may turn to the content.

- Spiritually, we can say that the Bible does develop an interactivity; because one is not content to receive the message, it is also necessary to question it, to put it into practice, to translate it into action, to visualize it, etc. The fact that the message of the Bible continuously refers back to Christ, and Christ refers us to the triune God – or on the contrary to humankind – isn't this a way of going beyond the text, of showing that the text alone, taken as a single literary work, is not in itself the word of God?

Christian theology, far from being an obstacle to the notion of hypertext, could on the contrary be a form of self-actual-

56

ization, in so far as it links the written word, the spiritual meaning, the human body and the invisible presence of God, in a way that means they cannot be separated.

The solitude of the individual and the personalization of faith

The Web can help to individualize faith and religious sentiment. We are tempted to construct our own religiosity privately, because the Internet allows each of us to communicate and to inform ourselves in an individual way: when we want, where we want, about whatever we want. Jean-Paul Willaime, director of the Centre of Comparative Sociology of Religions in Strasburg, emphasizes the link between the emergence of the Internet and the new religious context that favours individual emotions and personal feelings rather than a reflection founded on the dogmas and a rigorous analysis of the things on which faith is based. For Derrick de Kerkhove, director of the McLuhan programme on culture and technology at Toronto University and one of the most enthusiastic theorists about the new media, one of the characteristics of on-line speech is that it is intimate and private. Moreover, the Web makes possible individual control over public speech.

Alongside this individualization of faith, doubts are being expressed about the real quality of relationships in the virtual community. Under the heading "Is virtual integration a detached relationship?", Piero Bertolini of Bologna University identifies the two main problems of integration in the context of a virtual community. It is a question on the one hand of *relationship ethics*. How much is a relationship based on the values of an ethic of "deregulation", in so far as the partner in the exchange can change identity at will or hide behind a false identity? On the other hand, *the solidity* of the virtual link is called into doubt: What value is there in a virtual relationship that does not lead to a real relationship? Is an on-line community really a community, or is it a fantasy community? Without the support of a real live community, virtual communities risk disappearing as quickly as they appear. The more numerous and more general virtual relationships are, the less intense and significant they are. Virtual relationships are ambivalent, agrees Stephano Martelli, communication sociologist at Palermo University; they are the manifestation of the ageing of social relationships. On the one hand individuals withdraw from real territorial communities, on the other they regroup in other social circles, but without really feeling connected, or investing themselves with the same intensity as they would have done in traditional community relationships. The Internet adds to the phenomenon of uprooting the individual from his or her actual community context, amplifying a trend already well under way in the modern world.

Breton goes even further in his condemnation of the loss of social ties brought about by the new technologies. He entertains the possibility of a cause-and-effect link in the opposite direction. It is not these new religiosities that favour the Internet phenomenon, but the Internet, like other new information and communication technologies, that favours the new religiosities. These technologies want to create a religiosity founded on the particular needs of each person, outside of any community life. One of the widely recognized elements of the Christian faith – community life in an enlarged social framework – would be demolished by a new technology with esoteric and individualistic tendencies.

Thus there is conflict with the Christian ideal that those following the faith may do so only within the context of a community. The church is the gathering of those who have been baptized to hear the word. One of Jesus' first public acts is to call disciples to follow him (Mark 1:16-20; Matt. 4:18-22); after his resurrection, he appears to the disciples when they have gathered together (John 20:19-22), and his "ascension" occurs before a gathered community (Luke 24:50-51; Acts 1:4). From then on, the Christian faith has been a fellowship based on solidarity, in the image of the descriptions of the first church given to us in the Acts of the Apostles (Acts 2:42-47, 4:32-37).

58

Institutions are becoming obsolete

The Internet is a non-hierarchical medium, with no head. It knows no institution, hierarchy or delimitation. One Internet enthusiast says, "It is chaotic, anarchic, totally decentralized. No central administration manages it, nobody owns it, nor can anybody stop its development." Many people share that view. Castells would say that the first characteristic of virtual networks is "the value of horizontal, free communication", which is further strengthened by the other characteristic, which is the creation of a new form of freedom. Castells notes that in the United States in the 1980s it was in the marginal, anti-establishment and student environments that the Web developed most quickly. Furthermore, several writers have studied the American communities that were at the roots of the development of the Internet. One of them, the "hackers", is indeed a marginal academic world, which developed a form of ethics in which "defying authority and promoting decentralization" became an explicit aim. That community, which represented Californian counter-culture, thought about information technology and electronics as being "eco-warrior" or "hippy" technologies, and they believed that the messages exchanged ought to reflect the ideals of peace and harmony.

The Internet favours community ties without institutions or outside of institutions, where the mediators, web editors or moderators of forums – when there are any – are the only ones to have a formal role. Faced with a particularly "anarchic" development of virtual communities, there is a danger that the institutions end up being meaningless, or even being bypassed.

Flichy notes that this "community ideology" connected with the Internet is still strong, even though the way that the Internet is used is still a long way away from that initial utopia:

> The idea of a virtual group where everybody expresses themselves in an egalitarian manner continues to dominate. In these circumstances, the utopia becomes an ideology that partly masks the reality, but at the same time mobilizes the actors.

Flichy goes further still, because he studies the links between the Internet and politics, and shows that some people see the Internet as a model for a society without the state. He quotes one of the followers of this libertarian ideology, who says, "The Net promises us a new social, global and anti-hierarchical space in which anybody can, from anywhere and without fear, tell the rest of humanity what they think." For users of this kind, the Internet makes any control of information or any form of institutional mediation pointless. Some supporters of this ideology in the United States go so far as to want to see the disappearance of any state, or education system – institutions that are viewed as oppressive and anti-libertarian. This would make the Web anarchic in the political sense of the word. This has been called "cyber-selfishness". Flichy also quotes the British authors of a little book called *The Californian Ideology*, who condemn "the Californian ideologues who preach an anti-statist gospel of high-tech libertarianism: a bizarre mish-mash of hippy anarchism and economic liberalism beefed up with lots of technological determinism".

Is this philosophy an ideological by-product of the Internet, or is it one of its constituent elements? Both theories exist. Flichy, however, is in no doubt that the idea of virtual reality as a place without restrictions, making possible the creation of tomorrow's utopias, is a powerful breeding ground for new world-views based in the Internet.

The "horizontal" communication, free of hierarchy or constraint, that is developed by the Internet is not without its challenges. It is even destabilizing churches in the way they operate as institutions. There has been some reluctance to embrace the Web in Catholic circles, although the pope has spoken enthusiastically about the Internet's potential for evangelization and otherwise furthering the church's mission. However, some continue to see a threat to the church in the anarchic, entirely decentralized nature of the Internet. These warnings and concerns were widely expressed at a symposium on the new media organized by the conference of Italian bishops, at Assisi in 2000; some writers did not hesitate to talk about the Internet's "heretical" or "schismatic" tendencies. Catholics are not alone in voicing such concerns. An Internet that is too highly developed can be seen as real competition for the teams that run church institutions. It is important then to think through how church executives are to interact with the Internet. The Internet must be an aid to real communication in the church, but it should not replace those who are responsible for communication with the public. Because of the Internet, the church does run the risk of becoming marginalized or being sidelined. Because of its flexibility and its immaterial nature, the Internet is able to represent the universality of Christianity better than any institution, for institutions are necessarily limited in space and subject to the hazards of time. The Web is an effective and powerful tool. The danger of an "anarchic" use of the Internet makes us mindful of the fact that no technology is neutral in itself, and this one in particular is connected with a certain vision of the world, a certain ideological discourse, as well as with certain economic and commercial practices.

On the whole, the Internet remains – despite its concomitant temptations and limitations – an opportunity that the church must grasp.

NOTES

[1] Churches should be able to present a united front with all those who are fighting for a reintegration of the human into the economic, technological and ideological system that is establishing itself. The World Association for Christian Communication reflects on a communications ethic based on the values of the gospel. Its principles can, for the most part, be shared by all those who want a communication based on humanist values. Hence the document "The Christian principles of communication", available at *www.wacc.org.uk*

[2] Other people apply a less complimentary definition of a "hacker", as someone who seeks out the weaknesses of the system, a sort of information technology pirate.

[3] Cf. The exhibition "Les trois révolutions du livre" [The three revolutions of the book: parchment, printed book, e-book], Musée des arts et métiers [Museum of arts and manufactures], Paris, Oct. 2002-Jan. 2003.

4. The Internet and the Expansion of the Church

The church can neither ignore the phenomenon of the Internet nor be content just to criticize it, for both external and internal reasons.

Among the external reasons are that the church cannot allow itself to become marginalized or cut itself off from a population that is adopting the new information and communication technologies more and more: in March 2004, 45 percent of French people connected to the Internet, and there is every reason to think that this figure will continue to rise. After all, the church should go wherever people are, should answer the questions that people are asking themselves, should pay attention to what people are interested in, so as to be able to accompany them, warn them, educate them, direct them or help them to be able to find their own way.

The internal reason for the church to engage itself is that the Internet is a tool that can be of use in Christian witness and help to develop a participative culture so that church members can become more actively involved. The Internet is not just somewhere for displaying information: parallels are often drawn between the emergence of the new information communication technologies and the discovery of the printing press in the 16th century – sometimes to the extent of implicating Protestantism in both of these technological revolutions. There is sometimes said to be a link between Protestantism and the new information and communication technologies, just as there was between the printing press and the Reformation. The Internet is said to be a "Protestant" medium, while television is a "Catholic" medium. The Internet is democratic, individualistic and decentralized, whereas television is hierarchical, participative and centralized. Catholicism, which has people and places to showcase, is supposedly more comfortable with television, whereas Protestantism, which is often content with invisibility and with people who are more anonymous, is supposed to be more comfortable with a private and individualist media like the Internet. This line of reasoning has led one sociologist to say that "although, because of its more visible aspects, Catholicism had an advantage over Protestantism on television, Protestantism could find that it has the advantage on the Internet". This alleged difference between Christian traditions is relative, and it is often exaggerated. What *is* certain is that the Internet can contribute to the revival of a church that wants to progress further towards the wider participation of its members in its witness and its mission.

The dual reasons, social and ecclesiastical, why the church should welcome the Internet are expressed well by the direction put forward by the Italian Catholic church, in coverage of stages of meetings it organized about the new technologies:

> Cyberspace is not just an instrument or a channel through which people communicate, but a context in which the users themselves are able to change the physiognomy and the dynamics of communication. Within this world, with new territories and new languages, the church community has an active presence, in order to stay loyal to its mission of spreading the gospel.

To sum up, we can see three main ways in which the Internet may benefit the church.

1. The Internet as a tool for information. In this instance, the Web is nothing more than a sort of giant, virtual notice board that is continually updated. It is a matter of communicating as much information as possible, that is as up-to-date as possible, to as many people as possible, and of being able to reach the people who might be interested in that information in their homes. Users are not required to move, or to hunt for a newspaper or journal that they do not have at hand. This is currently the most common way of using the Internet, and also the most useful. This application poses very few problems, except that the information must be promptly and continuously updated.

2. The Internet as a tool for dialogue and exchange. This way of using the Internet is more complex, because it presupposes at least two partners in the exchange. This means venturing into a real process of interactivity, which is one of the most seductive aspects of the Internet yet one of the most complicated. Dominique Wolton recalls that communication is a subtle process that cannot be reduced to a simple technological exchange. This mode of communication is also one that may be used to turn outwards (external communication) or may be developed within the church (internal communication) because the Internet can operate simultaneously, both externally and internally. Here we are approaching a reality that is harder to define, because it is multi-faceted and is more directly connected to the nature of human beings themselves.

3. The Internet as a tool for enlightenment or a presence in the world. A tool for enlightenment: if used well, the Internet can be a formidable tool to allow believers to strengthen their faith and help them to tackle on a daily basis the spiritual questions that they have and the answers to those questions that help them to live their lives. A presence in the world: the Internet can very easily be used to address outsiders – those who are not part of the church but who might one day join it. The Internet can be a useful and effective tool for the expansion of the church, because although it addresses each of us, it also addresses all of us. The Internet is characterized by a dual use, individual and communal, both for existing members of the church and for all and sundry.

The Internet is indeed thought by many people in the church to be a means of enabling our communities to grow, of spreading the gospel, of bringing more people together, of reaching a different audience from those it usually addresses. Pierre Babin, a French-speaking theologian, has reflected on the relationship between the Internet and the church. His approach is rather too naively enthusiastic, but he is right to believe that the church would be marginalizing itself and would be disloyal to its mission if it did not engage more fully in the new media which are integral to a social revolution: "If the church were not on the Internet, it would be cutting itself off from the progression of history, it would be renouncing its spirit, which is the spirit of universal communication. By excluding itself from the Net, the church would seem, unlike Christ, to be refusing to give its body." And this Catholic practitioner suggests a definition of evangelization based on the most characteristic elements of the "new culture": in his view, it is not so much a question of using the new media as of becoming a means for communication. For the church, it is certainly this "evangelical" aspect that is the most innovative and the most fundamental.

Having identified five dangers that the Internet brings for Christian theology and practice, we would like now to put forward five areas in which it has the potential to bring about more effective sharing of the gospel and a renewed witness in the world.

In the service of local communities

Many community leaders who have a good command of the Internet have confirmed that the Internet can help in finding new active members, making communities more appealing, bringing people who are isolated into the fold. Thanks to the Web, the church can expand the area of its tent (Isa. 54:2). One minister claims that attendance at mass has doubled since he started producing and moderating an interactive parish Internet site. Another claims that he was able to increase his organizers and group leaders by 10 percent thanks to the Internet. It is no longer unusual for a minister, on asking new arrivals how they found out about the existence of the community or its activities, to hear the reply, "on the Internet".

Such new arrivals are numerous enough for us to wonder about their significance. It also means taking a gamble on the future by believing that what is happening now is an indication of other developments and is laying the groundwork for future growth. To illustrate this new side of things, we would like to quote the testimonials of two ministers and community leaders, one from Geneva and the other from Paris. Both of them actively use the Internet, not just as a practical tool for information, but as a forum for a new exchange, for a presence outside the church helping to make existing communities more animated.

Let us hear first of all from the Genevan Protestant minister Nils Phildius, who says that the Internet is helping his pastoral and parish work *(www.protestant.ch/direct/bernex)*.

This testimonial was offered at a seminar of the World Association for Christian Communication (WACC) on "Virtual Ethics in Europe: Challenges for Christian Communicators", which took place in March 2002 in Geneva. We have kept his testimony in its original format:

Who am I?

First of all, I am the minister of a rural Genevan parish...

I am an ordinary user of a very powerful information technology network that was set up by other people, but which I try to use to its maximum potential.

I am also the chairperson of the working group in charge of setting up and managing the Internet site of the Protestant Church of Geneva.

How is the Internet changing, or what does it bring to my ministry as a pastor?

In short: for me, the Internet is not just a way of communicating information (after all, information interests relatively few people...), but is a means of communicating a story – a narrative. It is this quality that makes the Internet a way of bearing witness.

Let me give you a few examples:

- Every month or so, and for almost nine months now, I have been sending a newsletter to many of the believers in my parish. The newsletter refers people to the parish Internet site, which includes an exhaustive list of all the activities of the parish. As a result, the number of families attending family services has almost doubled in six months! ...The newsletter may not be the only factor responsible for this achievement, but it has made a big difference. I am therefore convinced that the Internet can help to make parishioners more dedicated, give them a greater sense of belonging and communicate an enthusiasm to them; it really does help them, in my opinion, to be much more motivated about and interested in the life of the parish. I think that at the moment the Internet is the only means of communication that makes it possible to get these sort of results, because it is much faster and more flexible than any other.
- I try to put photos of parish events on our Internet site as soon as possible; people often tell me how sorry they were not to have been at such and such a festival, for example.

- We have a discussion forum on the parish website: a few months ago, a couple in my parish whom I didn't know started asking me theological questions via the forum, and through the questions and the answers, through this fascinating exchange, we started to build a good relationship. They now come to the service every Sunday. But they have admitted to me that they would never have dared to approach me directly: they were interested by the spiritual situation in the parish when they saw the website, and it was the confidentiality and anonymity of the Internet that gave them the courage to address a person of the church. Some people will say that exchanges in discussion forums will never replace face-to-face conversations – and I completely agree. But these forums are an excellent way of beginning a relationship, especially in a situation where there is often a great deal of mistrust of the church.
- I am shortly going to marry a young Hispano-Dutch couple currently living in London. They were at first very hesitant about getting married in church, because they were worried that the ceremony would be boring and joyless. But, as they are soon going to be moving to my village, they were curious enough to visit the parish Internet site. They liked the site so much that they sent me an e-mail to ask me to marry them, which I agreed to do because they are going to become my parishioners. We have exchanged several e-mails, and now I see them every couple of months, when they come to prepare for their move and their wedding, which will take place next September.
- For more than a year, my colleague and I have been posting all our sermons on the discussion forum; we quite regularly hear rumours of parishioners printing out the sermons – even though they have already heard them – to give them to their family to read, and so to share their discovery of biblical texts.

In short, although my parish Internet pages are of course visible throughout the world, I use them above all else as a way of communicating in my vicinity, and with the members of my parish. The website is becoming our common property; it tells the history of our community and highlights our activities by making events something for people to talk about. I have a fervent belief in the power of narrative: for me, our site narrates our community, giving it a story that is easily accessible to everybody.

Thanks to the Internet's speed and flexibility, the very story of our community becomes the witness, and sometimes the word, of life.

The second testimonial comes from Louis Pernot of the Reformed Church of l'Etoile in Paris. It is in the same vein as the previous one: when used properly, the Internet helps local communities to expand and grow, and is therefore a new tool for evangelization.

The Internet is a means of communication that is practically free and which has the potential to reach an almost infinite audience; it therefore represents an extraordinary opportunity for many people who have something they want to communicate, and that includes parishes.

But you can't communicate any old thing, any old how: the Internet is not magic, and "having an Internet site" won't mean that everything else will take care of itself.

An Internet site is a bit like a shop window: it only prompts any interest if it is placed in the path of people who might be interested. It is therefore important not to neglect the links that might cause people to pass by, nor to neglect referencing. And if the site is badly done, or not maintained – if, for example, the "spiritual life" page of a Website is always empty, or the "recent news" is two months old, or a discussion forum only has three boring questions – it can have a very negative impact on the image of those who are displaying it.

The task of setting up and maintaining a website must therefore be taken seriously, and not off-loaded onto the first budding enthusiast who comes along, or even to the first professional who introduces him- or herself, by leaving it up to them to do it. You have to remember that once a website is on-line, every month there will be hundreds of people who are in search of things that you can tell them, who will form an impression of who you are on the basis of that website...

So before producing a website there are a certain number of questions we have to ask ourselves: What do you want to communicate, to whom and in what form?

For a parish, it seems to me that there is more than one answer to those questions: a parish needs to communicate in order to live, to survive and to develop. Like any community, a parish needs a link between its members, to ensure that information moves between them and to inform them about what is happening: the Internet can do this very easily. A parish also needs to advertise itself to people who don't know about it and who might be interested: this is also something that the Internet can do. Then of course there might be the desire to evangelize, to spread the word and to bear witness to a certain interpretation of the gospel: the Internet makes this possible too.

Thus, the Internet can be a facility that allows parishioners to have the information they need to help them to participate more easily in the activities of their parish (Who is taking the service on Sunday? Who are the speakers at the meeting that evening? What time does it start? What time does Sunday school finish?), and to have that information immediately at hand, wherever they are (at work or elsewhere). That information is also more up to date than it is elsewhere. This facility can help them to become more involved in the church; and then you have to make others want to come... and to show them what you are, what you think, what you are doing. Experience shows that a website that is well constructed, when the necessary steps are taken to ensure that people visit it, can be extraordinarily effective. And it is effective in a quite unique way, because it tends to attract the young people of the parish, who are generally active and intellectually well-educated...

A parish is a community, and so the Internet must above all be something that leads eventually to direct contact. A website therefore needs to be used to attract people... and not to replace the parish by creating a sort of virtual community. I would therefore say that, paradoxically, a parish site should not be too complete: it should leave people wanting in some way (so let us stop short of broadcasting services on video, etc.) and should not try to communicate with the whole world, because a community is based on real relationships. The Internet has one big advantage, and that is that it allows the church and the gospel to go into people's homes; it is certainly much easier for somebody who is a long way away from the church to read something on an Internet site than for them to ask to be sent the printed parish newsletter. People have the impression of enjoying greater freedom, and of being able to find out more without having to get involved. But we have to make sure that it does not stop there; we must encourage Internet users to initiate communication in the other direction, at first by e-mail, and then invite them to make contact with the parish in person.

Lastly, it is important to remember that an Internet site is not set in stone. A website is only interesting if it is energetic, up to date, and interactive – and this takes work, and takes time. So before setting up a website, it is crucial to think about who will maintain the site, bring it up to date every week and see that it expands, and who will be responsible for maintaining search engine listings and links from other sites, etc.

Obviously, we have to communicate for our own sake, but also so as not to leave the field open to various sects or sectarian churches. But we must not communicate any old how.

Rebuilding community links

One of the main challenges facing the church today is living as a community and developing a sense of community among believers who are increasingly isolated, geographically and spiritually. The Internet can help to enhance – or to inhibit – a sense of community and to strengthen existing communities, perhaps even create new ones.

Community life, we know, is a *constitutive* element of the Christian faith, not an optional element. Christ did meet men and women individually, but he came above all to appoint disciples, in order to build with them a community of brotherhood, which was sometimes even in competition with the community of the family. Although that community disbanded for a while following the shock of the Lord's death, it quickly reconvened in the presence of Jesus Christ following his resurrection. It was then strengthened in the conviction that even when absent, the spirit of the resurrected Jesus Christ was always present in the community; a community that would not restrict itself to the original core of the first disciples, but would extend beyond the borders of Palestine to reach the pagan world, then expand to all inhabited land.[1] The missionary experiences of Paul showed that henceforth everybody could live in communion with Jesus Christ, and share this experience with others in the context of an active and united community life.

In this connection, it is interesting to read the greetings and salutations that open and close the letters of Paul. They show Paul's dual concern of addressing first of all a local community, a collective reader rather than a particular individual: "Paul... to the church of God that is in Corinth" (1 Cor. 1:1-2 and 2 Cor. 1:1). But Paul also addresses, beyond the local community, a larger, more universal community: "Paul, an apostle of Christ Jesus... to the church of God that is in Corinth, including all the saints throughout Achaia" (2 Cor. 1:1); "Paul... to those who are sanctified in Christ Jesus, called to be saints, together with all those who in every place call on the name of our Lord Jesus Christ" (1 Cor 1:1-2). It is difficult to imagine a more universal form of address. For Paul, the Christian community is both local and universal, exists both within a geographical location and without borders.

We have discussed earlier the danger of an exclusive and solitary use of the Internet, without however concluding that the Internet is intrinsically individualist. In fact, if anything it is the opposite that seems to hold sway today. The Internet makes it possible to recreate the social bond, nurture community links and maybe even develop new ones.

Contrary to the theory of Philippe Breton that the Internet constitutes a threat to the social bond, many testimonies and studies conclude that the Internet does not destroy the social bond but revitalizes it. Nonetheless, opinions differ on the question of whether the Web creates the social bond *directly,* by establishing real virtual communities, or only indirectly by prompting Internet users to turn towards real communities or to establish them. For others, the Web takes hold in an existing situation as an additional opportunity for communication, without profoundly changing the social landscape; it is simply an extra method of communication, with its advantages and its limitations: "The Internet risks weakening families where one family member isolates him- or herself to surf the net, but it also brings together families who are apart," says one Christian observer of the Internet phenomenon.

The Internet offers a novel mix of the local and the universal. It thus creates a new kind of community, where two contradictory dimensions blend together harmoniously.

Castells shows that a new kind of sociability is emerging that combines what is close by and what is far away. It is in this context of the new technologies that the new term "glocal" has emerged, a compound of the words "global" and "local". "Glocalize" is what the new media do – they blend and mix the global and the local, and thus transform our perception of the daily environment, says Castells. He has given a lot of thought to the Internet as a platform for a new mode of sociability that he calls "networked individualism". The on-line sense of community is in some way the culmination of an evolution that has already transformed the community bond, and that is now transforming the network into a virtual network. Networked individualism corresponds to the evolution of modern society, one of the dominant characteristics of which is the increased power accorded to individual feelings and attitudes; but this individualism comes once again within the scope of "personalized communities", of networks centred around the ego.

The Internet therefore creates sociability in two forms: on the one hand by maintaining strong links at a distance, thanks in particular to e-mail ("electronic mail also allows us to reveal our presence without entering into a deeper exchange, which we don't always have the strength for"); on the other hand by contributing to this new sociability built on networked individualism. Castells, who takes into account several sociological studies about the experiences of on-line communities, concludes, "Thanks to the flexibility and the power of the communication of the Internet, on-line social interaction is playing an increasing role in the social organization of society."

Pierre Lévy is one of those who think that the Internet can create true virtual communities, the communion and sense of belonging of which is as real – if not even more real – than in actual communities. In his view, the Web reconstructs a social bond based not on territorial affiliation, not on institutions, not on power relationships, but on other realities, such as an open process of collaboration, a sharing of power, common interests, play.

The Danish theologian Carsten Jensen believes that a virtual Christian community is possible, and it is to be considered a plus, in so far as it can offer something that a real community cannot offer: a church open 24 hours a day, present everywhere and visible everywhere. He claims enthusiastically that "the church on the Internet is democratic, dynamic and interactive". However, he does not go so far as to support the – extremely controversial – idea of on-line participation in the sacrament, of "virtual sacraments". That is the view of a member of a large Protestant church in Northern Europe that has long been the established state church, but does a French Christian, living in quite different circumstances, share his opinion? At least one Catholic activist is utterly convinced of the positive role that the Internet can play, not in creating communities but in revitalizing them: "Through the Internet, we are establishing characteristics of unity in the church. Christian Internet users see their church above all as a communion."

However, it seems a little premature to invest the field of virtual communities, or to claim at this moment that virtual Christian communities are possible and even desirable. We do not have enough objectivity, experience or evidence for that. It is possible that this is the direction we will go, but we cannot be certain. For the moment, the testimonies of communities that are entirely virtual seem too fragmented, and the communities themselves too fragile; they disappear as quickly as they appear, are characterized by the instantaneous and ephemeral and do not seem to develop into lasting relationships. As Patrice Flichy notes, networked individualism develops what he calls "weak links", in other words, interpersonal links which do not have an emotionally binding force. At some point, the virtual world must intersect with reality; an on-line community must encounter a real community that is geographically situated and physically present.

As yet, there are no entirely virtual communities – at least not in the churches. However, a number of virtual contacts can greatly enrich existing communities. This idea goes back to those who think that a virtual bond alone is not enough, and that it must lead to us finding a real bond, either with other people or those with whom we have met virtually. Many people emphasize the utopia of the notion of a uniquely "virtual community", yet believe nonetheless that real communities and on-line communities can be mutually reinforcing. This is the model that seems to be favoured by those who think in terms of church communities. Thus Kerkove, who is one of the most enthusiastic about virtual communication, writes, "The combination of traditional Christian community and on-line community makes us twice as strong" and Bruno Oudet, one of the earliest French proponents of the Internet, says, "The Internet will come to support communities, but will not replace meetings between Christians and their ministers."

In this way, the Internet allows us to experience a dual notion of community: universal and local, complete and restricted. The enquiries that we made in the communication service of the Reformed Church of France produced an interesting conclusion: in the case of French Protestantism, which is in a minority position, the Internet has shown itself to be a method that is particularly well suited to the way that the church is spread out geographically. When communities and believers live a long way away from one another, when parishes become more like regional churches than local ones, the Internet becomes an almost indispensable tool. In that situation, the Internet brings people together and creates a valuable sense of closeness in places where geographical distance separates and isolates.

It is certainly no coincidence that the Internet is most effective in two of the most scattered regions of the Reformed Church of France, where believers are very isolated from one another and from their community that is then no longer a "local" community. Where real communities are a long way away from one another, the Internet is a real help, a vital link, a link that can be indispensable for the very survival of communities often made up of activists who are motivated but isolated. The Internet allows people to develop close relationships, even truly neighbourly relationships, despite geographical separation. This observation, which is fundamentally true for a Protestant church in a minority position, might also apply to bigger churches, in view of the decline in the number of Christians in the main institutional churches in Western Europe. Roman Catholicism, at least in France, is also becoming more scattered.

Pierre Babin and Angela Zukowsky attempted to consider the local church – the traditional parish – and its relationship with the Internet. What does the Internet have to offer parishes? How does the Internet influence parishes? In their view, it is less a matter of moving towards new virtual parishes than of being aware of how the Internet will change existing communities: "Just as geographical parishes had to take on radio and TV in addition to the written press, so they will have to, little by little, take on new forms of media, relationships and groupings." They speculate that traditional parishes will evolve in four directions, under the influence of the new media:

1. *Towards an interactive and personalized support:* The interactivity of the new media and the emotional language of faith are called upon to blend together, if not to merge. The two central elements in this are the use of electronic media for services on the one hand, and the development of a personalized support on a daily basis (on the model of radio) on the other.

2. *Towards the growth in communities of affinity:* Parishes come to discover that they are based on a second foundation, "communities of affinity", which append themselves to traditional geographical roots. They are built on attractions and complementarities of communications. There are interesting analogies to be drawn between the media developing networked communities and the churches developing communities of affinity.

3. *The inner awakening and spiritual communities:* This entails sharing with others the conviction that Christ is the source and the point of one's life, but that conviction can be realized only in the sharing of the gospel with others. Each of us comes to learn that Christ is not a doctrine or something

unimportant, but the very foundation of my life and of all life. Babin would say, using a word that he happily borrows from the media domain, that it is a matter of sharing "the vibration of Christ in us".

4. *Priority to festivals:* All festivals have a considerable impact. According to Babin and Zukowsky, we must evangelize through festivals that have arisen from traditional Christian holidays. Churches must not hand over festivals to the media but reappropriate for themselves this means of communication and influence. The new media have a festive and playful side that we would be wrong to neglect out of concern for conveying the faith. We have seen in the testimonial of Nils Phildius that celebrating festivals and navigating the Web do have some reciprocal links.

Communicating better as a church

The church has been so concerned with helping others and spreading its message to as many people as possible that it has often neglected its internal, inner life, the way it functions in itself. But it is impossible to bear witness effectively on the outside without feeling loved, listened to and understood on the inside. The Psalms talk about these two forms of the word: first, the words of God to humanity, a clear and structured foundation for life and love. These are words that people avidly await, or else have difficulty hearing: "With open mouth I pant, because I long for your commandments. Turn to me and be gracious to me, as is your custom towards those who love your name" (Ps. 119:131-32). Second, the words that humans exchange between themselves and with which, collectively or individually, they address God. All these words are related, answer each other, intersect and often clash: "my lips will praise you" and also "the mouths of liars will be stopped" (Ps. 63:3 and 12), says the psalmist in the same composition.

Recently, the church has started to realize that it does not just have to communicate with others, but also with itself. Its first partner, in its work of listening and in its evangelical witness, is itself – its active and committed members, its employees, its officials and its ministers. How can there be any point in proclaiming a message externally if that message is not known and understood internally?

This problem is not new. From the beginnings of Christianity, one of the main concerns has been to manage to get people and communities who are very different from one another to live together and to communicate. The main tasks that occupied the first witnesses of Christ after his resurrection were to promote union and sharing, and to fight against separation and dispersion.

An overly hasty reading of the Bible, from the beginning of the Acts of the Apostles for example, could produce the impression that in the early years of the church everything was going very nicely. This impression is false, because we now know that the historiography of the Acts does not correspond to the reality; the accounts of an exemplary community life are too rosy to be true. On the contrary, from the outset Christian communities were troubled by all sorts of tensions (Acts 6:1-2), sometimes very serious ones. "But even if we or an angel from heaven should proclaim to you a gospel contrary to what we proclaimed to you, let that one be accursed!" (Gal. 1:8). It took mediators like Stephen (Acts 6:3 and 6), Timothy or Paul, to teach Christians to live together, to listen to one another, to accept one another and to accept others in their diversity: "If it is possible, so far as it depends on you, live peaceably with all" (Rom. 12:18), urges the apostle. Let us re-read the letters of the New Testament to see how huge and numerous the conflicts were; very quickly, conflicts of ideas or between people appear, factions emerge, debates about different customs cause problems, there are personality clashes.[2]

The many snippets of ethical advice given to members of these communities, and what is more to readers themselves, are also there to help us to accept each other, so that we will learn to communicate between ourselves and among all. The example of the dissension in the Pauline communities (Galatians, Corinthians) show us the extent to which poor communication – difficult relationships within communities and between communities – can confuse and weaken the message.

The Internet – especially in the form of an Intranet or Extranet (see appendix 1) – can be a remarkable tool to help us to communicate better, to help us accept each other in our differences, to develop a way of thinking and acting that is more community-based. The Internet can help us to communicate better as a church and within the church.

The Internet proves to be a particularly user-friendly and effective instrument that makes it possible to undertake projects together, to allow information to circulate, to reach those who are isolated, to fight against dispersal in time and space, to enjoy a community experience whilst being separated. The Internet allows us to communicate better as a church; according to various points of view:

- It is a tool that is particularly suited to the *dispersal of the church*, to churches in a minority position who are scattered over a large area. With the progression of secularization and the weakening of Christian communities (in terms of the number of active participants, either lay persons or clerics), it is the entire church that is becoming dispersed, and not just churches who have historically been in a minority position.

- It is a tool that, when used in the form of Intranet or Extranet, can greatly *help the internal communication of the church,* as is the case for every human institution (businesses and organizations). We learn that the Internet has come to play a more central role in information sharing, making it immediately accessible to as many people as possible. Without a policy for information sharing backed up by a participative use of the new media, the church runs the risk of drowning in a glut of messages that scarcely make it beyond the tiny circle of those who produced them.

- The Internet also makes it possible to fight against many attempts to keep hold of information, when some people want to defend their personal power or that of a small group to the detriment of a larger group. In this respect, it is a particularly democratic medium.

- The Internet makes it possible for us easily to build and consult *data banks* in order to preserve, add to and use the memory of an institution. This is particularly valuable for a church that wants to face up to the challenges of today's era and approach the future only with strict fidelity to the values that it defends, values that originate from the Lord and that it wants to convey as widely as possible to others. The Internet makes it possible to preserve and enrich the memory of the church.
- It is a powerful tool for coordinating activities, membership, managing everybody's diaries and calendars. It is involved in the organization and management of work, the distribution of tasks and the flow of work.
- Using the Internet to improve our communication as a church does, however, require a great deal of work by people beforehand in order to win the trust of partners and those in charge of the sectors affected (that is to say just about everybody), to reassure them, to explain how this new medium can be of use to them in developing their own activities. Without this prior trust, nothing would be possible. Because the Internet operates horizontally, there is a risk that it will be unsettling for some people who are used to a fundamentally hierarchical mode of operation. It would be naive to believe that, because Reformed Protestant ecclesiology is not hierarchical in nature, it is not hierarchical when it comes to the division of responsibilities. Various failures have shown the strength of the Internet precisely in that respect: there is a danger of it bypassing, even jeopardizing, some institutional modes of operation. On the other hand, it can also clarify them and make them easier and more transparent.

Intensifying and expressing one's faith

Many writers emphasize the individualistic side to the Internet. Unlike television, which is a community medium, the Internet is an individual medium. Philippe Quéau bases his research on that contrast: "The Internet essentially comes from the brains of the people who use it, whereas the richness of television depends on the financial means of those who control the programmes". But it is just one short step from an individual medium to a solitary medium. And we have seen that one of the risks of this new medium was to rupture the social bond (which is already very damaged) and to create "interactive solitude".

But this potential by-product does nothing to detract from the fact that the Internet, through its interactive aspects, is a medium that develops creativity and facilitates personal expression; it allows individuals direct access to a wealth of information; Internet users can choose the information, who should receive it, where the dialogue takes place, their own messages; they can even choose a (new) identity.

How can we define a little more precisely the link between the individual and faith, which would be strengthened by the new information and communication technologies?

We see three possible directions for the development of a confident and responsible faith with the help of the new media: theological education (deepening of faith); the expression of a joyous and lively faith (faith, joy and art); the intensification of personal feeling (inner life).

Thinking about faith

Every Christian is called upon to engage actively in the community and in the world. To do this Christians will have to be able to educate themselves in order to acquire a competence in a particular area, but they also have the mission of being able to work with others in the higher service of the church. Commitments in the church are generally group commitments which involve working and putting on projects with others, not by oneself. In these two areas, the Internet can be a valuable and effective aid.

An aid for training

Through the Web, Internet users can personalize their choices, choose where they get involved, freely direct their faith in one direction or another. Through the Web, Internet users become real actors in communication because they both send and receive. A pedagogical and didactic exchange is possible, allowing an educational relationship to be established through this new medium. This is something completely new compared to the conventional framework of communication that rested on the distinction between sender and receiver. Philippe Quéau contrasts the Internet, which is a tool for research and work (a societal tool), with television, which is essentially for entertainment and play; in his view, these are two communication tools, two totally different models. A major change has been introduced by these new media: the receiver also becomes a sender. The receiver can use this new medium freely, and personalize the way that he or she uses it in relation to his or her own needs. For Castells, free expression and personal creation are the most important of the latent requirements for the new communicators, because it is this that the traditional media cannot offer. The Internet is an opportunity for all of us to find on the Net the destination that interests us, to produce and disseminate our own information and create our own links.

The Internet can helpfully complement the existing educational bodies and faculties without replacing them. We will give two examples of theological training intiatives via the Internet:

- The faculty of theology of the University of Geneva has, since 1998, offered a branch of theological training by distance-learning, via the Internet (called *Formation @ distance*). The teaching comprises the transmission of texts, summaries of current publications, audio broadcasts of recorded lectures, the production of pages with frequently asked questions, with those documents being accompanied with set exercises, guided reading, homework assignments, forums and tests. The students may also talk to teachers at any time by e-mail. In terms of work and exams, the on-line student is subject to the same requirements as the students in the regular course. The study timetable provided for the whole course is divided into 19 "certificates", for which the student signs up separately. This system of certificates allows for a huge amount of flexibility, because it may be spread over eight years (as opposed to four for a regular student). On-line students must, however, turn up in person to the examination sessions. The studies are validated by the completion of 19 "certificates", which means 35 "semester modules" that the student must complete over the duration of his or her period of study. Every semester the student may choose to follow between three and five modules, with each module entailing between eight and fifteen hours of work a week. These 35 semester modules are available on the faculty Internet site, according to a predetermined timetable. Having chosen to follow one of the teaching modules on offer, the student is invited to log on regularly to the relevant pages, which offer a new programme of work every two weeks. For ancient languages, the student has not only the original text, with commentary and analysis, but can also hear it being read thanks to the "real audio" format. Classes and conferences may also be listened to on-line. The system used is the same as is used by radio stations to broadcast on the Internet. The texts are in the *html* format where there are hyperlinks, or *pdf* when it is desirable for the texts to be easy to print out; the sound is in *ram* format.

In 2003, there were 19 students registered on this distance-learning course, which for a small department is a suitable number. Among them, at least two did not live in Europe: one lived on the island of Patmos, and the other was a resident of India.

This first example of distance learning purports to be strictly academic. Our second example starts out with a different concern: offering basic theological training to as many people as possible.

- In late 2002, the Reformed Church of France opened another kind of on-line distance-learning project: *Théovie*. Open to all, with no prerequisites, and with no prospect of academic validation, this training offered a way of acquiring biblical and theological knowledge and invited people

82

to personal reflection on the major existential questions, based on the gospel. This programme aims to accompany those who subscribe to it in their spiritual quest, to help them to put themselves together, to give meaning and hope to their lives, to equip them so that they can experience the biblical message at the heart of contemporary history and culture, in the world and for the benefit of others.

This project aims to reach those who are for various reasons physically "remote" from the usual places and networks of biblical and theological training. It is also concerned with reaching everybody regardless of where they are and allowing personalized itineraries. This project is in its initial phase; when completed, around twenty modules will be available which will make up an organized and coherent course, which each of us may join at our own pace. They will be accessible by mail and on the Internet. The Internet makes it possible to develop original ways of learning based on interactivity and offers many possibilities for links (*e-learning*). A system of companionship and tutorials will be established with a view to personalized assistance. Sessions to bring learners together are also planned.

The worldwide nature of the Web provides potential for educational opportunities that will transcend traditional bounds of denomination and polity. In 2004, the Gordon-Conwell theological seminary in the US state of Massachusetts was collaborating with CentriHall Inc., a professional provider of e-learning services and content, to construct a programme in the Chinese language for churches in East Asia and elsewhere among dispersed Chinese-speaking communities. Courses were beginning to be offered in the Old and New Testaments, Christian theology, church history and the life of Jesus. The programme drew on an English-language e-learning curriculum already offered by the seminary students in 35 countries. If successful, administrators at Gordon-Conwell were hopeful that the Chinese programme of theological studies could provide a pattern for courses in such other languages as Korean and Russian.

We have briefly introduced several examples of relatively complete training, which is more or less self-sufficient. But Internet users may also choose to construct their own training course, simply by surfing the Internet at will. There are many sites offering Christian documentation and reflections, and we give some examples at the end of the book. The difficulty lies in making the right choices, and working without personalized follow-up. Although Internet users are free to choose among their acquaintances a person or a group who will help them to move from virtual learning to real dialogue, at some point a physical presence and a real meeting seem indispensable.

Building projects together

The Christian faith requires more than merely being intellectually understood, on the basis of careful study of the scriptures; it also has to be active. "But be doers of the word, and not merely hearers who deceive themselves," says James, who goes on, "For if any are hearers of the word and not doers, they are like those who look at themselves in a mirror" (James 1:22-23). We will see in this text that the biggest danger is that of withdrawal into oneself, expressed by a lack of visual communication because those who believe without doing are like those who look at themselves in a narcissistic tête-à-tête.

One of the challenges for the church today is to get diverse individuals to work together, because any ambitious project can be conceived and carried out only through teamwork. This requires a huge effort in terms of coordination, listening to one another and paying attention to everybody so that nobody is left out. At the same time, we must value the gifts that each of us has been given.

The Internet is a formidable tool for working together as a team, despite being physically a long way away from one another. It is a medium that brings people together. It may be used individually, but it also makes it possible to address several people simultaneously. In this way, it enables the building up, not of communities, but of *community projects*. Such projects cannot, however, be set up entirely remotely, or exist solely on a virtual basis. At some point, a physical meeting will be necessary. There is much that is analogical in communication – non-verbal language, what is not said, gestures, signals and facial expressions – that cannot be transmitted via the Web.

Faith, joy and art

The Christian faith must be intense and philosophical, but also playful and joyous. This is a vital element, introduced in Bible stories and sometimes experienced in the liturgy, but all too often forgotten in Christianity at the level of its philosophy and its practice.

The Ethopian whom Philip meets on the road to Gaza learns about the gospel and then asks to be baptized. He then "went on his way rejoicing" the text tells us (Acts 8:39). Christ "has given us a new birth into a living hope" it says again at the beginning of the first letter of Peter, which adds the order, "In this you rejoice!" (1 Pet. 1:6). Joy and faith need one another. The same goes for creativity. The fact of believing must make us creative, inventive and imaginative, because we are freed from the servitude of sin for a free life in faith. At several points, the Bible underlines that the believer moves from the world of shadows to the world of light ("him who called you out of the darkness into his marvellous light", 1 Pet. 2:10).

This joy in faith must, on the one hand, be expressed to everyone, which is the basis for a possible use of interactive media that turn us into transmitters and not just receivers. On the other hand, such a new medium, precisely because it is fundamentally interactive, makes it possible to develop the creativity in each of us. The Internet can thus open up a new space for freedom of expression and the exercise of faith. The Internet culture, Castells says, is a new form of freedom of expression: hypertext is personalized. Thanks to the Internet, we all can build our own systems of interpretation. This immediately raises the question of sharing common values, but at an initial level the Internet makes it possible to develop and strengthen the creativity of each of us, to integrate play and joy with research and self-expression.

It is therefore no surprise that artistic expression and the new media have come together very quickly. We have here two spheres of creativity that are mutually reinforcing and that call out to one another: some artists create virtually, using the infinite possibilities of the virtual language to create works of art of a new genre.[3] Software has been created specially for artistic purposes. Other artists use the new media to record their creations, when they are instantaneous, immovable and repeatable (installations, constructions, performances or *happenings*, etc.). There are clear similarities between the ephemeral nature of a work of performance art and the immateriality of virtual communication, which makes dialogue possible and easy. Many artists are able to make themselves known, but also express themselves and create, thanks to the new media. The Web has helped to reintegrate many artists into the social circuit, whereas television, through the influence of money, helped to marginalize them.

For Castells, there is more: art is not just a means of self-expression that is very easily "mediatized" through the Internet, it might make it possible to prevent the Internet culture sliding into the solitude of individualism. Art, thanks to the aesthetic experience, may act as a bridge between individualistic media and our fragmented culture. Indeed,

> art has always tried to bring together the various and contradictory faces of the human experience. More than ever, this could be its fundamental role in a culture defined by the fragmentation and not communication between the codes, a culture where the multiplicity of expressions could compromise the communion.

And Castells concludes, "Art, which is becoming increasingly a hybrid of virtual and physical materials, is thus in a position to build a cultural bridge between the Net and the ego." We live in a world of broken mirrors, made up of texts and initiatives that are not easily communicated, but art could restore the unity of the human experience across the diversity of experiences, forms and languages.

Have we moved away from faith? No, not if we consider that aesthetics now plays an increasing role in theological reflection, and this applies at all levels: exegesis, history of the church, homiletics, the search for new forms of the liturgy, catechisms, dialogues with culture and the inter-religious, etc. Faith must be able not only to dialogue with truth and justice, but also with beauty. It is therefore no surprise to see that a number of Christian Internet sites are constructed around artistic themes, if not dedicated entirely to this matter. Moreover, every Internet site begins by confronting Internet users with aesthetics, by presenting them with a welcome page that often has original graphics, some more successful than others. It is therefore no exaggeration to say that art is, literally as well as figuratively, a doorway to the virtual world.

Inner life

The Christian faith also means an intensification of this call of God in us, a call that we once received as a grace, a gift that was given freely and for free. No faith can exist without an inner dialogue within us and with God. Such an intensification of faith occurs every day, and sometimes involves a departure from the everyday for a period of meditation about or based on scriptures. The books of the New Testament encourage the internalization of this call. The language used expresses a profound truth: there is no faith without a personal – sometimes solitary – spirituality which internalizes faith and keeps it alive, like the inner and personal dialogue of Christ with his Father (John 17), as well as with his disciples (John 14-16). First Peter 2:4 shows us the direction to take, but also shows the situation of each of us before God: "Come to him, a living stone, though rejected by mortals yet chosen and precious in God's sight."

The spiritual experience can easily be communicated on the Web. Not that the Internet is a medium that is more spiritual than any other, nor that it has intrinsic qualities that automatically make it the vehicle for old and new spiritualities; we denounced the idea earlier on, along with others, that the very immateriality of the Internet might be the bearer of a particular spirituality. The Internet is nonetheless undeniably a channel that is particularly well suited to the individual intensification of faith, in connection with other personal research. Its fluidity and its speed, its simultaneously plastic and immaterial nature, and its interactivity, make it a favoured vehicle for spiritual expression. It is not a vehicle of faith, but a vehicle of its expression and of its experience. From there to a confusion of the two is only a short step that we must not take. Surfing the net is not praying, but the Web can be a channel that is particularly well suited for the practice of prayer, for experiencing a Christian spirituality in an interactive way.

What examples are there of a spirituality via the Internet? The Jesuits had the idea of using the Internet to experience more effectively with others the Ignatian spirituality. They offer a cyber-retreat, and invite Internet users to experience remotely the spiritual practices of Saint Ignatius Loyola. On the Website *www.ndweb.org*, an abbreviation of "Notre Dame du web" [Our Lady of the Web], the Jesuit Thierry Lamboley launches spiritual retreats on the Web that last three or four months, with an exchange of e-mails every three days, and discussion forums with the other participants. The site calls itself the Ignatian Spiritual Centre of Prayer on the Internet. This is a virtual place for a real spiritual experience. The organizer says,

> This way of navigating from site to site favours another way of approaching the real, which Saint Ignatius Loyola offers to help with prayer, by letting associations flow from a word, a prayer... One gambles here on the fact that the galaxy of meanings and comparisons, far from embarking on a fantasy that is disconnected from any reality, produces senses and gives access to a part of oneself in relation with one's history and with God. From the basis of something that seems worn out, multiple, transitory and heterogeneous, from what is sensed there can be produced an access to the permitted real.

He ventures to make a comparison between the Ignatian meditation technique and the computer technology that makes it possible to surf on the basis of a word, an idea, with a simple click. The effect of the Ignatian exercises on the Web seems conclusive: "It is not a virtual group, but a community that is formed," assures one "retreatee" at the end of a three-week forum. "The Internet makes it possible for remote believers to share their prayer and their reading of the Bible, by keeping regular appointments on the Web," says a participant. A further clarification: "Although the sacraments cannot be transmitted via communication technologies, they do permit some spiritual participation."

The German Lutheran church also readily uses the Web as a support for meditation, as a means that is particularly well suited to a personal intensification of faith and of prayer. One German theologian takes account of an on-line initiative in which several regional German churches took part at Easter 2000, called *www.webAndacht.de* [*Meditations on the Web*]. Among the various projects encouraging the use of the Web as a support for meditation, the authors study those linked to the Lutheran parish of Saint Elizabeth of Marbourg *(www.elisabethkirche.de)*. Under the heading "spirituality" there are four interesting offers: a "prayer chapel", the possibility of "meditations", a "prayer workshop", and a "mystic room". We are taking part in a process of establishing a system of management for the interactive possibilities of the virtual language and the spiritual reality of prayer, the virtuality of the one helping with the realization of the spirituality of the other. Thus, entering the "prayer chapel", the Internet user is first of all invited to light two virtual candles by means of a mouse click; then a book opens proposing various forms of prayers to read; one can also write one's own prayer on a virtual blank page and offer it to God and to the world. Once the prayer time is over, one is invited to extinguish the candles by clicking once again with the mouse. A virtual ritual is established for the benefit of prayer and meditation in real time.

Websites creating space for expression of spirituality are proliferating across the Internet. In the United States, the

United Church of Christ offers an assortment of resources for worship – including its own on-line prayer chapel – at *www.ucc.org/worship*. A second interactive feature offers a liturgy forum for pastors and lay leaders seeking deeper understanding of the church's worship life, the liturgical calendar and the biblical texts appointed by the Revised Common Lectionary. In addition to welcoming participants to the shared devotional life of the UCC prayer chapel, the worship page allows analysis and discussion of the various traditions of theology and practice that meet in this North American member of the "united and uniting churches" movement.

Does this herald a real Christian spirituality on the Net that is to come? It is doubtless too early to tell. In any case, it is interesting to observe that these virtual places for prayer are opening just as the places of worship are closing in the West due to lack of staff to run them and all kinds of material difficulties. The number and the variety of these initiatives concerning spirituality on the Internet[4] show that these churches that are afraid of leaving the domain of "spirituality" to others, have understood that they have to make use of possible connivance between mediation and meditation. The attraction of the Web and the need believers have to experience a spirituality in day to day life can meet harmoniously.

Building new ties

The church must testify to its message *in the whole world*, and this witness must be carried out with a sense of *solidarity towards those who are weaker*. Universality and solidarity are not optional requirements, but are fundamental for the church; they form part of its mission. These two requirements may be facilitated by the use of the Internet, on the condition that one can tell the difference between various ways of using this new communication technology.

Reaching as many people as possible

The Internet can help the church to fulfil its role of reaching as many people as possible. The role of the church is first of all not to guard the flame of a small number of followers, but to speak to the world, to speak to everyone. Christianity necessarily has a missionary vocation. It is not modelled on a closed community or a secret society but on an open community, visible to all, that addresses as many people as possible. That was the novel intuition of Paul of Tarsus, who brilliantly took up some of Jesus's attitudes and words: "You will be my witnesses in Jerusalem, in all Judea and Samaria, and to the ends of the earth" (Acts 1:8). According to Acts, these are the last words of the resurrected Christ to his disciples who were gathered in Jerusalem. Paul would take up again this universal vocation of Christianity, giving it a basis that is both theological and ecclesiastical. For those who have accepted Christ, he would say, "There is no longer Jew or Greek, there is no longer slave or free, there is no longer male and female; for all of you are one in Christ Jesus" (Gal. 3:27-28). In Christ, it is not just the chosen people, but also *all the gentiles* (Gal. 3:8). In saying this and arguing in that way, he lays the foundations of a religion with a universal calling, unconnected to a culture, a territory, a specific language, but to the single person and presence of the resurrected Jesus Christ.

Thus, the nature of the Internet corresponds to this universal ambition of the church of addressing everybody. The Web can allow the church to go outside of its four walls, to be present in the street, in the virtual public square. "The Internet opens the way to more than information sharing... Through the Internet, the church leaves the church and arrives on the square, in the underground and in offices, to the places where people live. Internet users straddle the borders... Their witness as Christians can receive a new audience," says one committed Christian Internet user. The Internet opens the way to an open and decompartmentalized world; it can offer a real forum for exchange on faith which will come to counterbalance an overly commercial use of this new communication technology.

One Christian user of the Net, Thierry Scholtes, notes that the Internet is changing our way of working, forcing us to expand our horizons in the projects that we undertake:

The Internet is a real forum for exchange... and also changes our way of working together: ever fewer meetings, ever more contacts tackle ideas before setting up a project. These electronic contacts make original meetings possible, by enlarging the field – including the geographical field – of our relations. The exchanges and the initiatives of the "core" Christians are multiplying: the network is also a support for a church that is first of all communion of those who have been *christened.*

If properly understood and used well, the Internet can allow us to become more universal in the way that we communicate, work and think in the church.

Fighting for a fairer world: a requirement of the gospel

One of the essential tasks of the church is to be particularly present among the weak, those forgotten by history, the poor and the powerless. This presence among neglected individuals also concerns peoples, countries and continents. Jesus lived and preached in the poor and neglected regions (Galilee) yet, by way of contrast, died in the capital city of Jerusalem, after having confronted the representatives of the (political, judicial, religious and economic) powers there. Although he addressed all without discrimination, he nonetheless favoured those who were excluded, the "little ones" (Matt. 10:42), the disinherited, the poor, those who do not even have any clothes to put on their backs (Matt. 25:31-46). According to the gospel of Luke, riches separate us from God (12:16-21,33-34, 16:19-31, 18:18-30) and material poverty can bring us closer to God (Luke 14:12-14,15-24, 16:19-31). Jesus' concern was not just to be close to those who were unfortunate, it was to talk to them, to communicate a message of peace, comfort and hope with a view to truly liberating them.

A passage from Luke connects this communication of the good news to the "little ones" with the arrival of the Holy Spirit, in a presence that might be described as virtual: "At that same hour Jesus rejoiced in the Holy Spirit and said, 'I thank you, Father, Lord of heaven and earth, because you have hidden these things from the wise and the intelligent and have revealed them to infants'" (10:21). Could an Internet that showed solidarity therefore be understood as a possible enlarged understanding of the role of the Holy Spirit in the world?

The church, which advocates a just and peaceful world, must find and make use of another way of using the Internet from the way that the commercial world offers us. In its endeavours to create a united community, built on the same values across the earth, the church should encourage an egalitarian vision of communication. On the other hand, it should denounce situations in which the Internet can be used purely for the profit of the wealthy, when it becomes a powerful instrument for the benefit of a single liberal economy.

Birth of the Internet and solidarity

A widespread idea and practice is that the Internet is an instrument for the benefit of the globalized liberal economy, which is to say for the benefit of the most powerful, of those who have the most money and knowledge. Market deregulation, the globalized economy and the world of the Internet are inter-related in many ways. The Internet is a medium that knows no borders, which is why it combines so well with a global economy, where all borders are abolished. This is an incontrovertible reality, but it does not have to be that way. The Internet is not inherently affiliated to a particular economic system. Several studies have shown that the Internet, which was born in the solidarity and unusual communities of students and scientists, could be to the benefit of a fairer development of society, with greater solidarity. We must be aware that there are several models of universality, and that the Internet can just as well be used to help the least developed countries grow as to exploit them.

We would do well to remember that the Internet was born in academic communities, in an environment that had similar ideals to that of the gospel: the free sharing of information, the establishment of a network of exchange other than the commercial work ethic. Thus, the Finnish academic Pekka Himanen has compared "the hacker ethic" of free and cost-free work to an ethic founded on financial viability and profit (which he calls "Protestant"!). He said that the "Net-economy" will fail precisely because it will encounter rejection or resistance to a globalization based on profit and the value of money. The economy and the sharing of *hackers* will develop a different concept of sharing, a different kind of globalization, based on "an immense and expert work which, like charitable and voluntary work, expects nothing in return except the desire to communicate, to act together, to socialize and to become differentiated, not through the exchange of services, but by 'sympathetic' relationships". This state of mind, which makes it possible to envisage a use of the Internet in a context other than that of the race for the biggest profit, does not fail to interest and challenge the church. This is the case, for example, with the Linux and GNU concepts, as opposed to those of the big multinational companies like Microsoft.

An example of ecumenical action in solidarity

It is significant that the joint charitable action of the Christian churches of Switzerland, Swiss Catholic Lenten Fund and Bread For All, chose "Sharing Communication" as the theme of its campaign for Lent 2002: communication, in the context of real North-South sharing. The material produced in the context of this campaign, designed for reflection by various churches, drew attention to global inequalities in communication, and to the commercial ideology related to this new industry, which earns an enormous profit from it. There is a paradox in this that we should be aware of:

> At the moment when new technologies make it possible for everyone and anyone to speak out, we are hearing only the big voice of the media giants who stifle all others. And the power of money scarcely shows any interest in the integration of those who are excluded.

In the context of this Swiss ecumenical action, the theologian Urs Jaeggi proposes some thoughts that delve into these observations in greater depth: he first of all denounces the mergers that take place between the giant communications conglomerates, which aim to control and dominate all technologies, programmes and information at a global level. This market logic brings about an unprecedented movement towards concentration.[5] At present, only around a dozen communications companies dominate the global market. Eight of the ten biggest conglomerates are based in the USA. This process of concentration and mergers also affects all the countries of the South. According to Jaeggi, this evolution does not apply just to technology and economics, but also concerns ideology, to the extent that it is guided by exclusive orientation towards profits. The maximum profit encroaches on values that are deemed to be secondary, like the well-being of everyone, social values and the notion of public service. Some very "political" issues like the digital divide, unequal distribution of wealth, the democratic principle of freedom of expression, jeopardized by the media power of those who have them, become unavoidable if we want to use the new technologies in an enlightened, ethical way.

In response to these global inequalities, reinforced by the new communication technologies, the ecumenical campaign for fair communication proposes a new order governing information and communication through the media.[6] The new technologies must be reconciled to this new vision of communication, for the benefit of sharing and redistributing of wealth, advancing particular cultures and the equality of all. The universality of the Internet and the universality[7] of the church can and must come together – conditioned on an ethically fair use of the Internet. The new communication technologies must not serve as a basis for a commercial and un-egalitarian society, but must help to build a more universally fair society, a society based on greater solidarity.

NOTES

[1] The biblical references are: Mark 1:16-20, 2:13-15, 3:13-19; Mark 3:31-35; Luke 24:13-21; Luke 24:33-43; John 20:19-29; Matt. 28:20b; Luke 24:50-54; Acts 1:6-10; Acts 2:36; Matt. 28:19-20; Acts 2:41.

[2] The biblical references are: Acts 2:42-47, 4:23-37; Gal. 1:6-10, 2:11-21; 2 Cor. 10-11; 1 Cor. 1:10-17; Acts 15; Rom. 14; 2 Cor. 2:5-11; Rom. 12:1-21, 14:13, Phil. 2:1-15; 1 Thess. 4 and 5:12-19.

[3] The preparations for the use of the new media by artists were laid with a method that is already old news in modern art: video art *(computer art)*, or art created on computers. One of the best examples of this trend is the American artist Bill Viola.

[4] We were really struck by the number – and the quality – of these virtual initiatives regarding prayer and spirituality.

[5] On 10 Jan. 2002 the largest global provider of Internet access, AOL, took over the world's number one media company, Time Warner.

[6] The Declaration of Manila, drafted and adopted in 1989 at the congress of the World Association for Christian Communication and completed six years later in Metepec (Mexico), proposes communication that balances out the powers, favours the development of humanity and avoids the levelling and the rendering uniform of culture *(www.wacc.org.uk)*.

[7] We use the word "universal" in the sense of global and egalitarian, in contrast to the word "global" or "globalization", which has acquired an inegalitarian connotation (hence we talk of a "globalized economy").

Conclusion

The Internet, if it is thought about properly and used well, can be of tremendous benefit to churches in their difficult mission at the dawn of the 21st century. Conversely, churches must be able to have a presence on the Net, at the very least to humanize it but also to tell the good news of Jesus Christ. Their mission, or one of them, is to help to make this new space for social dialogue into a place where authentically human exchanges are encouraged in an ethically fair manner.

But the churches' approach must go beyond this initial level of exchanges and shared interests. The emergence of this new virtual world is a chance for churches to live out the faith and put their ideals into practice, to submit their fundamental convictions to the test, to say who they are, what they are doing and on whose behalf they are doing it. Conversely, these information and communication technologies do not merely pose new questions of churches; they also remind them of the oldest questions.

We therefore propose four phases for this critical meeting of Christianity and the virtual world; four phases that move in the direction of the original and final meetings between the transcendence of God and the immanence of our surroundings.

Welcoming our world

The first phase recognizes the world as the site of Jesus' witness. We acknowledge our world as it is, and welcome the potential of what it is destined to become. This assumes an openness to modernity; a sense of curiosity; a mind for exploration; an analytical and non-judgmental mind, for our world is in flux. It is undergoing a metamorphosis, and it is changing fast. We have to learn how to accept this changing world, to understand it, know it and love it. We must do this in the name of the Christian faith that exists in our world today, not the lost world of yesterday nor the fantasy world of tomorrow.

A purely abstract approach does not belong to the hermeneutical tradition of the Christian faith. The truth cannot be separated from its historical context, nor the expression of faith from factual history. The credo of apostolic faith places a historical point of reference at the heart of its expression of fundamental truth: "He suffered *under Pontius Pilate.*" History, in Christian perspective, is first of all the history that we experience, illuminated by the witness of the scriptures through the inspiration of the Holy Spirit.

Under the influence of new technologies, there are a number of indications that our society is indeed evolving. The following is a random selection of some signs of these changes, that are connected with the evolution of technologies and the establishment of global communication networks.

- The replacement of time lived with the notion of real time: communication becomes instantaneous.
- The relativization of geographical place; distance is no longer an obstacle to exchanges; the place where we communicate scarcely has any impact any longer; the emergence of the virtual, which is presented as a new form of reality (or a new way of living in reality); interactivity, which allows unlimited dialogue between users through the mediation of a machine.
- A new form of writing that mixes forms of expression that were once separated or even opposed: text, image, sound, graphic form, movement; more technical ideas such as hypertexts, which make it possible to go from one text to another, one site to another, or one place to another without having to change place or environment; new letters are appearing, like @, the "at" symbol, which has become the de facto 27th letter of the alphabet.
- A new means of communication, electronic mail, which combines the advantages of post and the telephone, etc.
- A communication which is becoming asynchronous (it is not necessary to be present continuously in order to be able to communicate).

Before wondering where the gospel fits into this context, we must take the time to discover this new world, to explore it, to question it, and to fear it for what it is, what it can give us and what it might take from us.

Hearing the word

The main task of the church is to listen to the word and communicate it to others. But how do the new communications media fit into this special mission that is Christianity's? Do they fall under the category of listening to the word or communicating it? If these media are called upon to become the word, there is a risk of the message being replaced by the technology. If we classify them under communication, there is then a risk of losing sight of the pure purpose of the Christian message: a church that merely communicates will no longer be the church of Jesus Christ. Contrary to what is widely believed today, we do not hold that communication justifies everything or sanctifies the message it is conveying. The purpose of the church is not to communicate, but to make Jesus present, to create the conditions for a meeting between humanity and the One given to the world by God.

In addition to communication and exchange, there is a form of directed message we shall call "speech". It is regarding this key concept of speech that we now reflect. Here we content ourselves with specifying and differentiating between four different forms of words which meet and mingle in the communication of the Christian message. It is necessary to distinguish between these forms in order to put the new technologies at the service of the communication of a true word.

- *Oral or spoken dialogue* (what Philippe Breton calls "oratory"). This certainly plays a part in the communicative process, but not all communication necessarily comprises speech (animals communicate but do not speak). Speech in the primary sense of the word presupposes a distancing between the world and oneself; it makes it possible to act, to effect a change (in ourselves, in the world); it can serve to reduce the violence of society.
- *The internal voice, inner monologue.* Some words elude the process of communication: these are more ancient and profound words, which do not develop a specific purpose. These words run deeper than any communication (whether that communication is channelled through tech-

nology or not). These are words that presuppose and express something internal. Words are the basis of an inner dialogue – that unique ability that humans have to converse with themselves. Before communicating with others, we have to be able to nurture our own internal dialogue.

- *The scriptures become the word.* In Christianity, scripture takes centre stage. However, these texts represent only one moment in the process of communicating God's message to humankind: a record of oral speech, first of all, which is destined to become living speech again. This is a word that mingles with others to tell the history and current events of a human race in search of God. The word of God and the words of humanity are therefore inextricably linked. These mingled words can take various forms: initial texts and their commentaries, words spoken or thought, signs and symbolic gestures, body language and images.

- *The voice of the new media.* Lastly, there is speech channelled through the media, old or new. A new, different kind of speech made up of a mixture of sounds, words, pictures, surroundings, signs. Pierre Babin, who has experience in working with communication technologies and has given them a great deal of thought, calls this form of speech the "ground". The "ground" comprises the context of the mediatized speech, its structure, the foundation, the background. This term, he tells us, expresses a global intuition, a mood, the background of consciousness. The "ground" is silently carried along as the context of the speech; it is therefore a sort of anti-speech, counter-speech and non-speech. And yet the "ground" talks, talks to us and talks to our contemporaries. You will note that the Internet consists nonetheless of text, essentially written, but this impression is ultimately false, because the writing is so labile, so ephemeral that it is closer to being speech.

Perhaps we are heading towards a new form of speech, one which is no longer spoken but is comprised of virtual writing in real time. A communication technology expert would joke that when writing becomes increasingly interactive we will have reinvented the telephone! Observations about virtual writing that is tending to become speech pose a challenge for the church, one of whose main tasks is to turn a written text (the Bible) into living speech.

We are not trying to construct a theory of communication in a Christian context – we simply want to show that the concept of "speech" corresponds to a variety of realities that are sometimes contradictory, and that they need to be reconciled to one another. The old problem of human speech in contrast to the word of God finds itself enriched within new parametres, due to the evolution of the philosophy of language and to the arrival of new technologies.

Unmasking the idol

We have talked about the risks and distractions and the lure of cultural idolatry that come with using new technologies. The "idol" is lurking everywhere.[1] It has a virtual presence in every network, the technologies, the thought systems, whether they are religious, material or communicational. We therefore have to be constantly on the look-out for the idol's presence, often hidden behind its opposite (for example, the "religiosity" of the Internet, which we have seen was, in fact, a form of paganism, in the service of an unavowed materialism).

There is nothing fundamentally new in this. First of all, the "masters of suspicion", analysts like Marx, Nietzsche, Freud identified idols in human systems, thought and motivation. We learned to drive out the idol from its deepest hiding places, because it hides in human beings, in their words, their actions or their subconscious. But the new media restore the idol's power as they employ seductive, powerful and universal technologies. The language of advertising encourages us to see technology as a source of human development. Many adverts for software or computer programmes show us the consumer as Adam or Eve, giving in to the temptation to pick the forbidden fruit: the limits have been crossed, the prohibition superseded, the contract broken. Humankind is portrayed in search of its divinization, aspiring to become the equal of its Creator, the glorified human being is exactly what is offered by some of the discourse about the new technologies. On occasion, the technologies themselves induce this vision of humankind.

We must continue to ask critical questions, in the three areas affected by the new media; these are the areas of humanity (anthropology), society (sociology) and Christianity (theology). For the time being, without any pretence of being exhaustive and in a slightly anarchic fashion, we simply wish to identify ten places in which the idol might be present in connection with the new media. We have here ten areas in which we are aware that we must exercise caution, which form a sort of cautionary ten commandments for a computerized world.

1. It is too easy to succumb to the fascination of technology, to deify it; humanity too easily finds itself in the service of new gods.

2. It is easy to fall into the trap of communicating for the sake of communicating, which means to communicate without purpose; not giving communication a higher goal comes down to sanctifying communication itself.

3. One is easily captivated by the immediacy of communication in real time, no longer giving oneself time for reflection.

4. Beware the temptation of self-worship: expressing yourself may become more important than receiving information from others, talking more important than listening. Do not put yourself before others, becoming greedy for resemblances (looking for your own attitudes in others) and mistrusting differences.

5. Beware of seeking efficacy above all else. Strength (of the new technologies, of money, of power relations) thrives due to the weakness of humanity, our own self-contradictions and those of the world.

6. Do not give precedence to appearances and to the external over the depth of your being; beware of not heeding your own inner voice.

7. Refrain from amplifying the momentary, giving it a status of being something more exceptional than it is. Each moment is a part of the process of being formed. Do not under-estimate time, the structuring dimension of history (individual history as well as collective history).

8. Beware of losing sight of an enduring goal. Living for the moment, it becomes difficult to make choices, because everything seems equally worthwhile. One lives without bearings, without higher values. One no longer recognizes moral priorities; one no longer gives priority to defending ideals.

9. Do not let the whole take precedence over the specificity of individuality. Otherwise, history may be reduced to recurring atemporal cycles. The individual may seem but an element of one, big Whole which either surpasses the individual or occasionally elevates it. The individual is thus either magnified in our thinking, or wiped out entirely.

10. Refrain from dividing the concept of human being into body and spirit, and the world into the material and the spiritual. This creates dualism and confusion. One confuses the human and the machine, substance and thought, the human (without a body) and the divine (without a heart).

Seeking the image of God

Having identified the idol, it remains for us to seek the image of God – the ultimate aim of the Christian quest of faith. This quest for the image of God – an old Calvinian theme – can be articulated only in the outward manifestation of the Holy Spirit in the world.[2] But how much can any image depict, and why, in a religion based on the word, is the image so important? We identify three systems in which we may discern the image.

- Christianity is fundamentally a religion of the image, as much as of speech. There is no Christianity without image, no revelation without incarnation, no transmission without mediation. And that is true even in the Protestant system and in the context of a theology that comes from the most iconoclastic branch of the Reformation. Pierre Gisel noted,

A theology that would forget or underestimate the reality of the image inscribed in all writing would simply have lost its field of work and application... Let us not be mistaken. There is an imaginative reconfiguration... in the apprehension of the real, and not simply description.

- Asking oneself about the human also involves asking oneself about its image, both the image of God (Gen. 1:26-27) and the images, internal or external, which it produces. The spiritual experience of Christianity is, we know, realized in the carnal. Just as at the centre of the new media there is a complex of issues regarding the nature of human beings, their bodies and the images, real or fantasy, that they produce or project. This anthropological question is central to any consideration of the new media. As the Jesuit Luc Pareydt affirms, "It is possible to be joined again, yet differently, to the time of the new information and communication technologies."
- Lastly, the new technologies are producers of images, of new images, frightening or seductive, mental or material: logos, the graphics of writing, animated clip art of gifs, photos and video sequences... This sundry and varied collection of visual elements that we call, for want of a better word, "images", form part of "writing" in the new media. Different sorts of images, as well as writing or the word, make up technological communication.

But with these images, we still find ourselves in the realm of symbolic and human representations; we do not attain the image of God, the pattern for our faith. Can we go further, and draw nearer to this first, original image for want of being able to imagine it? Can the new communication technologies which sometimes invoke the idol bring us closer to the image of the One who eludes any representation? We think it possible to evoke three elements that point in this direction:

1. We have to reaffirm that the visibility of God is ahead of us and is not within our control. It is ahead of us, beyond the horizon of understanding and perception. But at the same time, this image of God continues to be promised to us, and we know – we believe – that one day it will be granted to us. This is a quick sketch of the field of Christian expectation, the future we are heading towards, our eschatology. New technologies will not be able to replace this expectation, as we have said several times, but they can stimulate it, reactivate it, in that they also dwell in the domain of utopia. There are of course differences in the nature and strong tensions between the technological utopia of the Internet and an expectation of the promise to be fulfilled in the kingdom, but there are also some connections to exploit.

2. Traces of the image of God are present in the world. God does not leave us orphaned. Christ was given to us as the perfect image of God. This image of God is living in the celebration and the action of the churches and the faithful. But other traces of this image of God also live on in the world, in a form that is more symbolic than sacramental. As soon as we believe that the world can be the theatre of the glory of God (Calvin) and not just a stage for mankind's disasters and massacres, this conviction must also extend to the virtual world. There is in fact no reason to stop simply with the "real", as long as one accepts – as we have argued – that there can exist a *true* virtuality that is real, and not just a virtuality that is no more than a deceptive and truncated reality.

3. The most elaborate expression of the image of God may finally be found in the Christian practice of faith: in prayer, reading and meditation on the scriptures, liturgy and celebration of worship, actions and expressions of solidarity. Can such a spiritual life, based on prayer and praise, be accompanied or even helped by the new media? It is doubtless too early to answer that question definitively. We have witnessed and reported some happy experiences that move us towards an affirmative answer, and we will continue to study them. For the moment, we remain open to initiatives that make it possible for as many people as possible, including

those from outside the churches, to discover through a variety of means and media the infinite riches of God through the love of Jesus Christ.

NOTES

1 Humanity being inclined to evil, it risks creating idols everywhere, all the time, and in different circumstances. Jean Calvin, in his *Institutes of the Christian Religion*, I,XI,3 challenges any anti-Jewish tendency in the church on that basis, in the name of the idolatry that threatens us all: "this people with fervid swiftness repeatedly rushed forth to seek out idols for themselves as waters from a great wellspring gush out with violent force. From this fact let us learn how greatly our nature inclines towards idolatry, rather than, by charging the Jews with being guilty of the common failing, we under vain enticements to sin, sleep the sleep of death". (Calvin: Institutes of the Christian Religion I, Library of Christian Classics, ed. and trans. by John T. McNeill and Ford Lewis Battles, Philadelphia, Westminster, 1960).

2 Despite the reputation of Calvinism for presenting a plain outward aspect, John Calvin himself was emphatic in regard to the *visual* dimension of faith. In *Institutes of the Christian Religion*, IV.xiv.6, he recalls that Augustine defined "a sacrament" as "'a visible word' for the reason that it represents God's promises as painted in a picture and sets them before our sight, portrayed graphically and in the manner of images". In *Inst.* I.vii.5, he writes that the scriptures reveal God to believers "just as if we were gazing upon the majesty of God". Another example of this visual imagery is Calvin's description, in *Inst.* I.ix.3, of the work of the Holy Spirit "who causes us to contemplate God's face".

Appendix 1

Laying the Groundwork for an Internet Project

PROJECT PROTOTYPES

An Internet project should begin with two questions: (1) Who is communicating with whom? (2) What kind of information is being exchanged? A number of media solutions are available, depending on the answers to those questions.

An Intranet: This is generally an internal network, which means that only members of the organization have access to it. Basically, an Intranet is set up within a network that is inaccessible from outside, but it is also possible to make it accessible remotely by requiring the user to enter a password.

An Intranet, which presupposes Web pages, should not be confused with a local network, which consists of sharing the disks and printers through a network (although an Intranet can of course use a local network as a platform).

> The World Council of Churches has an Intranet that is accessible only to employees both from inside the internal computer network or from outside via Internet. Via the Intranet, records of meetings, information about departments, and all kinds of practical information is made available to everybody who works there. The Protestant Church of Geneva also has an Intranet network that can be accessed remotely via the Internet.

An Extranet: This involves opening up the organization's network to an organization's external partners. Those partners are defined and identified, and have access only by means of a user name and a password (e.g. *www.ecuspace.net*).

> The World Council of Churches is using a system of this kind in order to allow identified ecumenical partners to access information pages, to communicate and to exchange documents.

The *Internet:* This is when Web pages are made available to the general public. This kind of access does not usually require users to identify themselves. An Internet site is in theory more diverse because it addresses a wide range of different audiences, each of whom should find something there that interests them.

INTERACTIVITY

Another way of looking at the different ways of using the Internet is to analyze the level of interactivity of the exchange. Let us take four typical examples:

		Destination	
		A single user or site	Multiple users or sites
Source	**Single user**	E-mail applies in this instance; exchanges are sporadic in nature.	Websites fit into this category; an Internet user can move easily from one page to another, but the pages themselves stay frozen. If the pages are not updated frequently, the user is likely to lose interest and not return.
	Multiple users	An example of this would be the gathering of information via the Internet, such as via an on-line form. Commercial sites might also fit in here.	Discussion groups come in here. This also applies to work groups or websites where several Internet users contribute to the content.

In any Internet project, it is important to decide on the appropriate level of interactivity: a high level of interactivity requires considerable technical and human resources in order to regulate an environment that is constantly changing, but too little interactivity will deter Internet users and consign the website to being dismissed as uninteresting.

THE FIVE STAGES IN SETTING UP AN INTERNET PROJECT

An Internet project is a bit like editing a newspaper, in that it can be done by one person in an evening or may involve a year's work for ten people. Three things are decisive for the success of the project:

The purpose and the content of the website: This means answering the questions: Who wants to communicate with whom? What do they want to say? The answer will determine the choice of a technological solution.

The management of the website: Given the potential impact of a website, good organization is imperative in order to guarantee the quality of the information presented. The contributions of those providing the information to post on the site will need to be managed, and the content and format will have to be agreed. In the case of a discussion group, where everybody can express him- or herself freely, debates will need to be moderated in order to make sure that they remain in keeping with the spirit of open debate.

The technical infrastructure: The network and the computers are of course a factor for success. One question that will arise is whether to purchase your own organization's technology and computer equipment or to contract a specialized company to host your website for you so that all the technological factors are taken care of.

Taking a professional approach to setting up an Internet project involves the following five stages:

Stage 1: Preliminary review
- Conduct a preliminary review: Why the Internet? What makes the Internet important for your organization? Have we got enough resources to get started? Who should be involved in the project from day one? Are we talking about internal or external communication?

Stage 2: Feasibility study and concept development
- Study the existing situation and define the communication objectives.
- Design a plan for implementation: the objective of the project and its framework; the participants; their roles and responsibilities; the planned activities and the expected result; the organization of the project; the funding; the risks.
- Conduct a detailed study, together with all those involved, of what will be needed for this system. What changes will each of the participants see as a result of the new system?
- Develop the operational specifications. What capacities will the new system need? How will it be organized?
- Develop the non-operational specifications. Will the system need to be available 24 hours a day? What security provisions, such as anti-virus protection or back-up solutions, will be needed?
- Develop a prototype in order to test the specifications. Are we sure that we understand the specifications properly? Are they realistic?

Stage 3: The search for a solution
- Define, research and then choose technology and management solutions: What products will be needed? Who will be responsible for setting up the technology? Who will manage the technology system?
- Issue an invitation to tender, search for a provider or search for a partnership.

Stage 4: Implementation
- Build the technological solutions.
- Organize how they are going to be used. Who will do the backing up, provide the technical support, etc.?
- Train users in relation to the role they will play.
- Carry out tests to check that everything works the way it should and to correct errors.
- Validate and give final approval before the launch.
- Launch.

110

Stage 5: Maintenance and website moderation
- Ensure technical maintenance of the website.
- Manage and moderate the website.

Although this checklist will of course need to be adjusted to reflect the scale of the project, it shows that setting up an Internet project is more about communication and organization than about the technology that is needed – it would be a mistake to develop an infrastructure without first thinking properly about content and organization.

IMPLEMENTATION BY STAGES

In order not to be overwhelming, Internet projects are usually implemented in a number of small, successive stages. This step-by-step approach allows an Internet project to be launched quickly and affordably, and also means that each stage is an opportunity to test the key elements of the following stage. Because the world of the Internet is both new and complex, any Internet project is a min field of potential pitfalls arising from an inadequate grasp of the real implications: taking things one step at a time means that there is room for adjustments to be made as necessary.

PROJECT ORGANIZATION

Organizational skills are at least as important to the success of a communications project as technical knowledge.

The community

If a website is to thrive, it has to become a forum for exchange for a real community. An understanding of that community, its expectations and its way of life is therefore crucial in order to be able to design a website that will suit it. Editorial and graphics decisions and the main functions of the site all flow from an understanding of the community. The question of target audience is central: that means knowing who will access the site (the public, identified users, specific groups such as ministers, church officials, etc.)

Management of the website

Management of the website is fundamental, since it will be the basis for the organization of the whole project. Who will do what? How will the information be checked, with varying degrees of flexibility in relation to the information type? Who will be able to add information (i.e., will contributions be centralized and/or decentralized)? For example, it is important to distinguish between information that represents the official voice of the organization and contributions made in a personal capacity by individuals. An editorial team will need to be trained to run the website in the long term.

The communications officer

The communications officer will need to put in place a communications strategy; the Internet project will be merely the platform for that strategy.

A website is designed in the same way as a newspaper. The communications officer will have to design an external communications strategy, adopt an editorial stance and make choices about graphics, the plan of the site and navigation logic. If the website is interactive it will be particularly important for somebody to moderate debates and to process the messages received.

In the case of an Intranet, the communications officer will need to define and put in place a strategy for information sharing within the organization; he or she will need to have an understanding of who will provide information and who will receive it, and of how the information will be transmitted.

In the case of an Internet site, the communications officer will need to be familiar with the mass media; for an Intranet he or she will need organizational management and group motivation skills. It is important to stress that the role of the communication officer is primarily editorial, and not technical: the classic mistake made by many churches is to give the task of setting up their site to a brilliant young computer whizz, without having given any thought to the content – producing nothing but an empty shell.

The project leader

The project leader has responsibility for the practical implementation of the project. It is important to stick to the stages of implementation, despite the temptation to skip over them in order to save time – for example, by developing a solution on the basis of a vague idea, without having defined what is really needed, without having considered several possible solutions and without consulting future users, or hurrying through the preliminary tests so that mistakes have to be corrected after things are already up and running.

The project leader has the vital task of ensuring that enthusiasm for the project is not allowed to detract from a professional and methodical approach to its implementation; the project leader also has to help overcome people's doubts or reluctance, to mediate conflicts and to ensure that everybody's efforts come together in a coordinated fashion and progress at the same pace.

Choosing a technological solution

You might think that choosing a technological solution would be straightforward (you just ask a technician to do it), but there is much more to it than seems at first. A website is like a living environment: choosing a technology partner is the beginning of a long-term relationship, and it is therefore important to make sure that you will get along.

One of the other decisions to make is whether or not to have your own in-house Internet server: there are many companies that can house a website, allowing the organization to concentrate on the structure and content of the site, and not on problems with the server or with telecommunications.

It can also be a good idea for organizations to collaborate when choosing a technological solution, because two entirely different websites can coexist in the same technological environment. Federated organizations have a particularly important role in facilitating technological collaboration; what is more, practical collaboration of this kind can create an enhanced sense of mutual assistance and improve relations between organizations.

It is also important to consider the implications of the choice of technological solution for those who will use the site. For example, imagine what it will be like for people in Africa – or even in a remote corner of Switzerland – who are using a slow Internet connection via a telephone line: you will need to design pages without too much content, without too many pictures or pictures that are too big, and with little or no sophisticated "flash" animations, so that it does not take forever to download pages. Do not forget to estimate running costs for the medium and long term, being careful not to underestimate the cost in terms of working time and any hidden costs and the difficulty in evaluating precisely the complexity, time and cost of an IT project.

Typical Websites

The websites listed below have been selected because they illustrate different typical ways of using this new medium in the context of the church. Their presence here does not imply particular approval by the authors; the aim and quality of some are in fact questionable. It is, however, interesting to analyze them to see the variety of ways new media are approached.

Christian portals, resources and search engines

These websites are entry-points to finding Christian websites. Very often they cover one linguistic area, and sometimes they are linked to a group of churches. They offer mainly links and resources with key-word search and/or thematic search possibilities.

www.crossdaily.com (English) is a directory which claims to provide links to 20'000 Christian websites. It also offers online Bible search, possibility to meet other Christians, e-cards and other entertaining Christian resources, and a shopping facility.

www.ekd.de (English, French, German) is the portal of the Protestant churches in Germany. In addition to a large number of links, it contains a variety of resources like e-cards and contacts.

www.gospelcom.net (English) offers access to Christian resources through the search engine crosssearch.com and a Bible search via biblegateway.com. It also proposes services like the verse of the day, newsletters and shopping facilities.

www.protestants.org (French) is a portal to French Protestantism. Maintained by the Protestant Federation of France, it carries websites of a large number of Protestant churches and organizations, and public statements on present political and social issues.

stjrussianorthodox.com/stjrusorthodoxy.htm, run by the Saint John the Baptist Russian Orthodox Church in Rahaway, New Jersey, USA, has published a complete directory of links to Eastern Orthodoxy websites.

www.museeprotestant.org, a virtual museum that offers an opportunity to explore the history of Protestantism, especially in France, a country with an old Protestant tradition. This website will soon be available in English and German as well.

Interaction

These websites provide the possibility of interacting with the church by asking questions, or participating in forums or training sessions.

www.losung.de (German and 17 languages) is run by the Moravian Church in Germany and offers the famous "Watchwords" *(Losungen)*. Each day a Bible verse in chosen randomly from either Testament. This practice has existed since 1728!

www.protestant.ch (French) is the website of the Protestant churches in the French-speaking part of Switzerland. It offers access to services such as, for example, *www.questiondieu.com*, an interactive question-and-answer tool for the wider public, and *www.protestant.ch/liturgie.nsf*, a database for liturgical resources.

www.sacredspace.ie (English) is run by the Irish Jesuits to "invite you to make a 'sacred space' in your day, and spend ten minutes, praying here and now, as you sit at your computer, with the help of on-screen guidance and scripture chosen specially every day".

www.questiondieu.com is a place where anybody can ask a question about religion. A Reformed theologian provides an answer. What started as a limited initiative in one church becomes now a collaboration between churches in two countries (Switzerland and France).

Virtual presence

The following are examples of what a virtual presence can mean practically. It is particularly interesting to see various models of linking the virtual and the real dimension of each initiative. Some are more an extension of an actual physical organization, some others exist only in the "virtual world".

thecyberchurch.org/ (English). This virtual church is animated by Pastor Darril Beaton from the Friendship Baptist Church in Litchfield, Connecticut, USA. Many services are offered to the "virtual" members such as on-line worship services, scheduled on-line teaching, e-mail notification of events, chat rooms, possibility to submit a prayer to the prayer team.

www.cyber-church.com/ (English). This virtual church (formerly known as Cyber Church of Eufaula, Oklahoma, USA) claims to be one of the very first truly on-line church ministries. It is a ministry of Christ Ring Ministries and offers many services, such as a prayer room and discussion forum.

www.godweb.org/ (English). This church claims to be "the first church of cyberspace" and was founded by the Central Presbyterian Church of Montclair, New Jersey, USA, where Charles Henderson, the website's animator, was pastor until 1996. The site even proposes a virtual sanctuary with a proposal of sermons, music and prayers. You can also see some further reflections from "Presbyterian Today Online" on this topic at the address *www.pcusa.org/today/archive/features/feat9811h.htm*

www.goarch.org/ (English). The Greek Orthodox Archidiocese of America, based in New York, offers many resources (daily readings, prayers, newsletters), as do many other churches. In addition, it provides a live weekly broadcast of the divine service and it is possible to visit the chapel using virtual reality.

www.ecic.info/ (English). This website of the European Christian Internet Conference does not offer particularly specific "virtual" features, but it is worth mentioning that this organization has no physical home (as it is a network of people and not an institution); its only visible home is on the web (and at annual meetings).

www.ndweb.org/ (French). This website, run by an association based in Versailles (France), considers itself an Ignatian spiritual centre (from Ignatius Loyola, founder of the Jesuits). It offers a wide range of possibilities: a prayer to the prayer team, participation in continuous prayer, and even a spiritual retreat on internet guided by e-mails, with the possibility of contacting some experienced members of the Catholic church if needed.

www.partenia.org (French, English, German, Italian, Spanish, Portuguese, Dutch). This is the website of the "virtual diocese" of the former Catholic bishop of Evreux in France, Jacques Gaillot. Here, cyber reality and reality merge: Mgr Gaillot was dismissed from his position and put in charge of the diocese of Partenia, which no longer exists because this land is now covered by the North African desert. Jacques Gaillot has found a clever way of going around the difficulty by establishing a "real" presence in the virtual world.

* * *

Ecumenical organization websites
The *Directory of the WCC* (annual) lists known websites of its member churches and of national councils of churches, Christian world communions, and international ecumenical organizations. More websites can be found at the address *www.wcc-coe.org/wcc/links/index-e.html* maintained by the WCC web office.

The World Council of Churches
www.wcc-coe.org is the institutional web site of the World Council of Churches. It provides access to information about the WCC's activities and organization. It also contains a list of links to many ecumenical players as well as a press corner for journalists.

Apart from the main site, the WCC manages the following sites, with specific objectives:

www.wcc-usa.org and *www.wcc-un.org* are the sites respectively of the WCC's office in the USA and the UN liaison office, both in New York.

www.overcomingviolence.org features an interactive site for churches and groups to share events, stories and documents concerning the Decade to Overcome Violence (2001-10).

www.eappi.org, the site of the Ecumenical Accompaniment Programme in Palestine and Israel, carries stories written by accompaniers on the ground, and a rich photo selection.

www.photooikoumene.org provides a collection of professional quality photos on the ecumenical movement. More historical photos can be found in the WCC's archives on archives.wcc-coe.org.

Onlineservices.wcc-coe.org is a central registration site for signing up to receive by e-mail information and documents on the WCC's events and programmes.

Appendix 3
General Resources

For a historical account of the creation of the Internet and an introduction to it, see:

Abbate, Janet, *Inventing the Internet: Inside Technology*, Cambridge MA, MIT Press, 2000

Berners-Lee, Tim, *Weaving the Web: The Original Design and Ultimate Destiny of the World Wide Web*, New York, HarperBusiness, 2000

Flichy, Patrice, *L'imaginaire d'Internet*, Paris, La Découverte, 2001

Guedon, Jean-Claude, *La planète cyber. Internet et le cyberespace*, Paris, Gallimard, 1996

Huitema, Christian, *Et Dieu créa Internet*, Paris, Eyrolles, 1996.

The **first writer who invented cyberspace and virtual reality** was a science-fiction writer, William Gibson, in his novel *Neuromancer*, 1984.

For the **Roman Catholic Church**, see "L'Eglise aborde ce nouveau média avec réalisme et confiance" (The church approaches this new medium with realism and confidence), message from Pope Jean-Paul II of 24 Jan. 2002. Some major Catholic websites include: *www.vatican.va*; *www.cef.fr*; *www.croire.com*; *www.ndweb.org*. Cf. also the proceedings of the Florence seminars organized by the Catholic Institute of Paris and the Catholic Faculty of Central Italy in Sept. 2000: *Ratio Imaginis. Expérience théologique. Expérience esthétique*, published in *Vivens Homo*, 12, Florence, Jan.-June 2001.

For the office of social communication of the Italian bishops conference, see *www.chiesainrete.org*

Artists exhibiting and creating on the Net with a religious approach (which may or may not be visible in their art): *www.c-vonaesch.ch*; *www.al.for.com*; *www.gouillyfrossard.org*; and for photos of art: *www.merzeau.net*

On-line distance-learning:
– Reformed Church of France project: *www.theovie.org*
– University of Geneva, *Formation @ distance* site: *www.unige.ch/theologie/distance*

Appendix 4
Reading List

The authors draw from the following works. Exact references appear in the French edition of this book.

Arnold, Matthieu, "Annoncer l'Evangile: les réformateurs et le médium écrit", in Richard Gossin ed., *Eglise.com*, pp.55-69.

Babin, Pierre and Zukowsky, Angela Ann, *Médias, chance pour l'Evangile,* Paris, P. Lethielleux, 2000.

Barbrook, Richard and Cameron, Andy, *The Californian Ideology*, 1995. *http://www.alamut.com/subj/ideologies/ pessimism96942945960* 479613lidIdeo-I.html

Barlow, John Perry, "The Great Work", *Communications of the ACM*, vol. 35, no. 1, Jan. 1992, pp.25-28.

Barlow, John Perry, "@ Home on the Ranch" paper at the @ home conference, doors of perception 4-6 Nov. 1994, Amsterdam.

Bertolini, Piero, *Chiesa in rete*, www.chiesainrete.org, p.31.

Blaser, Klauspeter, *La théologie au XXe siècle, Histoire-Défis-Enjeux,* Lausanne, L'Age d'Homme, 1995.

Bobert-Stützel, Sabine, "The Medium Is the Message?", *www.theomag.de, Magazin für Theologie und Ästhetik,* vol. 7.

Castells, Manuel, *The Internet Galaxy: Reflections on the Internet, Business and Society*, London, Oxford Univ. Press, 2001.

De Kerchove, Derrick, "La Parole on-line: e-vangelizza-zione", *Chiesa in rete*, p.19.

de Rosnay, Joel, *L'homme symbiotique*, Paris, Seuil, 1995.

Debord, Guy, *La société du spectacle*, Paris, Gallimard, 1992 (1967).

Debord, Guy, *Commentaires sur la société du spectacle*, Paris, Gallimard, 1992 (1988).

Demissy, Claude, "L'Eglise, le 'net' et la catéchèse", Richard Gossin ed., *Croire.com*, p.107.

Diot, François, "Pour bien vivre l'Internet", *Fête et Saisons* no. 555: *L'Internet, du nouveau dans l'Eglise*, May 2001.

du Charlat, Régine ed., *L'art, un enjeu pour la foi*, Paris, Editions de l'atelier, 2002.

122

Finkielkraut, Alain and Soriano, Patrick, *Internet, l'inquié-tante extase*, Paris, Mille et une nuit, 2001.

Flichy, Patrice, *L'imaginaire d'Internet*, Paris, La découverte, 2001, p.85.

Gisel, Pierre, *Croyance incarnée*, Genève, Labor et Fides, 1986.

Hermann, Jörg and Mertin, Andreas, "Virtuelle Religion. Die Herausforderung der neuen Medien für Theologie und Kirche", *www.theomag.de* , vol. 7.

Himanen, Pekka, *L'Ethique hacker*, Paris, Exils, 2001.

Hoog, Emmanuel, "Internet a-t-il une mémoire?", *Le Monde*, 17 Aug. 2002.

Jensen, Carsten Riis, summary in English of his thesis on the Internet and Christianity (in Danish), see *www.ekklesia.dk/speciale-engelsk*

Kauffmann, Roland, "Adam et Prométhée: la question des techniques de communications aujourd'hui", Richard Gossin ed., *Eglise.com*, p.75.

Kreisberg, Jennifer Cobb, "A Globe, Clothing Itself with a Brain", in *Wired*, June 1995, *www.wired.com/wired/archive/3.06/teilhard_pr.html*.

Lamboley, Thierry, "L'impact du virtuel: vraie ou fausse communication?", *Christus*, 189, 2001, *www.ndweb.org*.

Le Monde, "L'éthique *hacker*, nouveau paradigme social", 3 May 2002.

Lévy, Pierre, "Internet, un monde sans porte?", *Les cahiers protestants*, Feb. 2000.

Lévy, Pierre, *World philosophie*, Paris, Odile Jacob, 2000.

Martelli, Stéphano, "La società mutata: nuove identità, nuove relazioni", *Chiesa in rete*, p.31.

McLuhan, Marshall and Fiore, Quentin, *The Medium Is the Massage: An Inventory of Effects*, New York, Bentham, 1967.

O'Grady, Ron ed., *Christ for All People: Celebrating a World of Christian Art*, Geneva, WCC Publications, 2001.

Pareydt, Luc, "Effet de miroir ou effet de sens? Faire corps au temps des images. Brèves variations anthropologiques", *Un monde en réseaux. Enjeux et défis de la com-*

munication à l'heure d'Internet, le 20 septembre 2002, Paris, Centre Sèvres.

Ponnau, Dominique, *Figures de Dieu. La Bible dans l'art*, Paris, Textuel, 1999.

Quéau, Philippe, "Interview: 'les méta-mondes' permettent d'explorer de nouveaux contenus", on the website of the INA for *Imagina* 97.

Scholtes, Thierry, "Ce que propose Internet aux croyants", *Fête & Saisons*, 555, p.23.

Schopfer, Olivier, "L'Eglise protestante de Genève sur Internet", *La vie protestante*, 224, March 1998.

Willaime, Jean-Paul, "Les incidences de la 'révolution' communicationnelle sur la vie religieuse", Richard Gossin ed., *Croire.com.*, pp.34-35.

Wolton, Dominique, "L'Eglise face à la révolution de la communication et à la construction de l'Europe" (The Church in face of the communication revolution and the construction of Europe), in P. Bréchon and J.P. Willaime eds, *Médias et religion en miroir*, Paris, PUF, 2000.

Wolton, Dominique, *Internet, et après? Une théorie critique des nouveaux médias* [The Internet: what now? A critical theory of the new media], Paris, Flammarion, 2000, p.101.

Wolton, Dominique, *Penser la communication*, Paris, Flammarion, 1997.

Zahrnt, Heinz, *Aux prises avec Dieu. La théologie protestante au XXe siècle*, Paris, Cerf, 1969.

Zukowsky, Angela Ann, "Un nuovo senso del luogo per l'evangelizzazione: l'era virtuale e il Vangelo", *Chiesa in rete*.